Lukas Gerhard

Magnetoelectric coupling at metal surfaces

Experimental Condensed Matter Physics
Band 7

Herausgeber
Physikalisches Institut

Prof. Dr. Hilbert von Löhneysen
Prof. Dr. Alexey Ustinov
Prof. Dr. Georg Weiß
Prof. Dr. Wulf Wulfhekel

Eine Übersicht über alle bisher in dieser Schriftenreihe erschienenen Bände
finden Sie am Ende des Buchs.

Magnetoelectric coupling at metal surfaces

by
Lukas Gerhard

Dissertation, Karlsruher Institut für Technologie (KIT)
Fakultät für Physik
Tag der mündlichen Prüfung: 20. Juli 2012
Referenten: Prof. Dr. W. Wulfhekel, Prof. Dr. G. Weiß

Impressum

Karlsruher Institut für Technologie (KIT)
KIT Scientific Publishing
Straße am Forum 2
D-76131 Karlsruhe
www.ksp.kit.edu

KIT – Universität des Landes Baden-Württemberg und
nationales Forschungszentrum in der Helmholtz-Gemeinschaft

KIT Scientific Publishing 2013
Print on Demand

ISSN 2191-9925
ISBN 978-3-7315-0063-6

Contents

1 Introduction

Nowadays, the driving force for technological advances in nano-magnetism is the ongoing demand for magnetic mass data storage devices. After Oberlin Smith had proposed to use magnetic particles as information storage in 1888, it took about half a century until IBM created the first computers using hard disks. As is well known, magnetic memory technology experienced an unprecedented rise which is still not finished. Although new developments took place on the field of reading heads which culminated in the application of large room temperature TMR sensors [1, 2], the working principle of how a magnetic bit is written basically has not changed: a current flows through a coil, thus magnetizing its ferromagnetic core which in turn magnetizes the bit to be written. Following this technique, new hard disks with higher storage density were developed by simply scaling down the dimensions of all parts. But there are certain fundamental limitations of this technique which recently lead to a slow-down of the increase of areal storage density. Although several attempts have been made to find new ways of information storage such as MRAM, FLASH and SSD, the vast majority of information is still stored on ordinary hard disks. Conventional magnetic bits are composed of several hundreds of magnetic grains. This high number ensures a high signal-to-noise ratio [3] which scales roughly proportionally to $\sqrt{N}$. Thus a reduction of the bit size has to go along with a reduction of the grain volume V. This, in turn, reduces the thermal stability of the orientation of the magnetization which is given by the magnetic anisotropy energy E_A. Once this energy barrier becomes too small, thermal energy $k_B T$ may lead

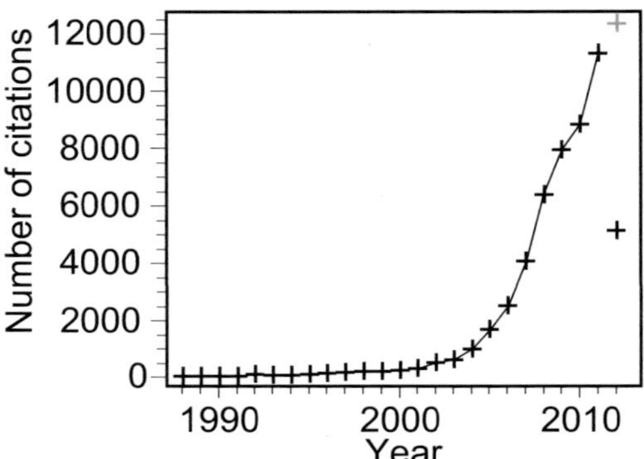

Figure 1.1: Citations of publications including the key word 'magneto-electric' (after Web of Science ®)

to a reversal of the magnetic bit and thus to a loss of information. Therefore, on the one hand, the desired lifetime for storage devices of about 20 years requires $E_A = K \cdot V > 60 \cdot k_B T$ (where K is the anisotropy constant and k_B is the Boltzmann constant). On the other hand, an increase of the anisotropy K leads to unwritable magnetic bits: higher anisotropies require higher magnetic fields to switch the magnetization of the bit. The magnetic materials used in today's hard disks, however, are already the best that we know. This contradiction of readability, stability and writability is known as the magnetic recording trilemma. Although the limitations described by this simple estimate have already been overcome, the basic ideas still apply. Minor modifications of this technology, e.g. heat assisted recording which lowers the anisotropy by local heating during the writing process, only delay the advent of a fundamental new approach. In order to avoid the projected failure of Moore's Law of an exponential increase in integration density, new technologies are required in the near future.

The possibility to change the magnetic state of matter in a

controlled way by means other than a magnetic field attracted significant interest in the last years (see Fig. 1.1). Magnetoelectric coupling (MEC) allows the magnetic state of a material to be changed by an applied electric field. The ultimate dream is to use MEC in magnetic data storage devices to write magnetic bits and to circumvent the issues summarized above as the magnetic recording trilemma. To date, MEC has mainly been observed in insulating materials such as complex multiferroic oxides where ferroelectric and ferromagnetic order parameters coexist [4, 5, 6]. The microscopic processes that lead to MEC in single-phase compounds are rather complex and still lack a complete theoretical understanding. In a different approach, multiphase compounds are engineered with the ferroelectric and ferromagnetic phases in close contact so that electric and magnetic dipoles couple via the interface [7, 8]. In both approaches, the electric field changes the electric polarization by displacing positive and negative ions with respect to each other. This displacement in turn influences the magnetic state.

Since the discovery of the first multiferroic materials more than forty years ago, only little attention has been payed to MEC in metals. As in metals any external electric field is screened by the formation of a surface charge, MEC cannot be found in a bulk metallic system. But this surface screening charge extends into the topmost atomic layer and leads to slight displacements of the positively charged atomic cores. This effect was observed for the first time in nanoporous non-magnetic platinum [9]. A metallic nanostructure whose magnetic configuration changes upon these lattice relaxations on the picometer scale could open the possibility of MEC in metals. The idea of electric-field-induced magnetization reversal in metal thin films was theoretically first described by Blügel in 2000 [10]. The first experimental evidence for an influence of an electric field on the magnetism of a metal thin film was given by Weishaupt *et al.* [11]. In ultrathin FePt films, they found a change of the coercivity of 4.5 % under the application of electric fields. A Japanese group found large changes in the anisotropy of magnetic thin films in TMR junctions under application of high electric fields [12]. A bistable toggle switching of the magnetization direction between two bistable orientations has been reported very recently by the same group [13]. Against this background, we used a scanning tunneling microscope (STM) to study MEC and were able to demonstrate for the first time bistable switching of the magnetic order on the nanoscale. An STM is an ideal tool with which to investigate this because it can image surface structures on the nanoscale and it provides high electric fields underneath the tip to induce phase transitions via MEC. As model system, we chose two atomic layers thick Fe islands grown on Cu(111): Fe is known to have a structural instability between face-centered cubic and body-centered cubic phases as well as a strong variation of the magnetic order following slight changes in the unit cell volume [14]. For this model system we have shown that nonvolatile magnetic information can be written on the nanometer scale by the application of high electric fields. The effect of piezoelectric forces exerted on metal-surface atoms by a high electric field is not limited to the particular system of Fe/Cu(111): basically, any phase transition of first order in a thin film that is prone to subtle changes of the surface lattice could be triggered by an electric field. This opens up a huge playground for experiments on MEC in metallic systems. Since metal thin films have been subject to theoretical studies for a long time, these comparatively simple systems could allow a thorough understanding of the underlying physics. It is of fundamental interest to understand why magnetic properties which were usually regarded as intrinsic properties of a material can be changed by electric fields. Possible applications are not limited to the field of data storage devices, but can also be found in the wide area of spintronics.

This introduction is followed by a brief description of the magnetic phenomena encountered throughout this thesis, the theoretical background of STM and an overview of MEC. A third chapter describes the experimental setup that was used. The chapters four to six report the experimental results on MEC in 2 ML Fe/Cu(111), 1 ML Fe/Ni(111) and 10 ML Ni/Cu(111). The final conclusion summarizes and discusses the obtained results. The theoretical calculations presented side by side with the experimental results in chapter four were performed by the groups of Arthur Ernst from the Max-Planck-Institut in Halle and Ingrid Mertig from Martin-Luther-Universität of Halle.

2 Fundamentals

2.1 Magnetism in thin films

In chapter 6, the influence of an electric field on the spin reorientation transition (SRT) in Ni/Cu(100) thin films will be discussed. Therefore, a brief introduction of magnetic anisotropy of metallic thin films and the two experimental techniques that were used for studying magnetism in these films will be presented here.

2.1.1 Magnetic anisotropy and spin reorientation transition

The energy needed to magnetize a ferromagnet depends on the direction of the magnetization, where the energetically favored direction is called easy magnetization axis. This effect is described by the magnetic anisotropy. In small single-domain ferromagnetic particles (e.g. magnetic bits), the magnetic anisotropy hinders the rotation of the magnetization from one direction along the easy axis to the opposite direction (which is equal in energy). The magnetic moment of transition-metal thin films mainly stems from the spin moments of the electrons. The easy magnetization axis of such thin films is determined by the following contributions to the total magnetic anisotropy:
Magnetocrystalline anisotropy arises from the energy difference between different alignments of the spin moment via the orbital magnetic moment which feels the crystal lattice. In general, this is a rather weak contribution of the order of a few μeV per atom

in the bulk. In a cubic lattice, the magnetocrystalline anisotropy energy per atom is to second order given by

$$u_{mc} = K_1(\alpha_1^2\alpha_2^2 + \alpha_1^2\alpha_3^2 + \alpha_2^2\alpha_3^2) + K_2\alpha_1^2\alpha_2^2\alpha_3^2, \qquad (2.1)$$

with α_i being the cosine of the angle between the magnetization and the i-axis and the anisotropy constants K_1 and K_2.

Furthermore, most thin films are strained due to the lattice mismatch with the substrate. This reduction of symmetry can lead to a considerable contribution to the magnetic anisotropy, called magnetoelastic anisotropy. Although the strain and thus the magnetoelastic anisotropy may vary with film thickness, this contribution also scales with the volume of the film. The related energy per atom for a cubic lattice is [15]

$$u_{me} = B_1(\epsilon_{11}\alpha_1^2 + \epsilon_{22}\alpha_2^2 + \epsilon_{33}\alpha_3^2) + 2B_2(\epsilon_{12}\alpha_1\alpha_2 + \epsilon_{23}\alpha_2\alpha_3 + \epsilon_{13}\alpha_1\alpha_3),$$
$$(2.2)$$

where ϵ_{ij} is the strain tensor and $B_{1,2}$ the constants of magnetoelastic coupling.

As thin films can usually be considered as a single magnetic domain, the energy related to the magnetic fields is given by

$$u_{ms} = \frac{1}{2}\mu_0 M^2 cos^2\theta, \qquad (2.3)$$

where μ_0 is the vacuum permeability, M the magnetization and θ the angle between the magnetization direction and the normal of the film plane. Minimization of this stray field energy, also referred to as shape anisotropy, rotates the magnetization into the film plane.

Besides these volume-dependent terms, the reduced symmetry at the interface and the surface leads to an anisotropy energy $K_s + K_{int}$ proportional to the film area and independent of the film thickness. For comparison with the bulk contributions, it has to be divided by the thickness t:

$$u_s = \frac{K_s + K_{int}}{t}.$$
(2.4)

The combination of these contributions may lead to different orientations of the easy magnetization direction depending on the film thickness, lattice mismatch and surface morphology. Furthermore, the anisotropy constants are also temperature-dependent. Typically, thin ferromagnetic films undergo a transition from out-of-plane to in-plane easy magnetization direction with increasing thickness. This SRT can be explained by a dominating magnetoelastic anisotropy for very thin films that are strongly strained and thus show out-of-plane easy magnetization direction. In thicker films, the strain is partially released by the formation of dislocations and thus the shape anisotropy becomes dominating and rotates the easy axis of magnetization into the film plane. In chapter 6, we will see that in the case of Ni/Cu(100) the surface anisotropy may induce a second SRT.

2.1.2 Hysteresis loops obtained with the magneto-optic Kerr effect

A rather simple method to measure changes in the magnetization of thin films is the magneto-optic Kerr effect (MOKE). It was first discovered by J. Kerr that polarized light reflected from a magnetic surface changes its ellipticity depending on the magnetization of the surface. In the experiments presented later in this thesis, the polar MOKE configuration was used: the magnetic field is applied perpendicularly to the surface and a polarized laser beam is incident at a small angle with respect to the plane normal (see Fig. 2.1a)). In this configuration, both the change of the ellipticity and the rotation of the polarization of the reflected beam are approximately linear to the relative change of the magnetization component perpendicular

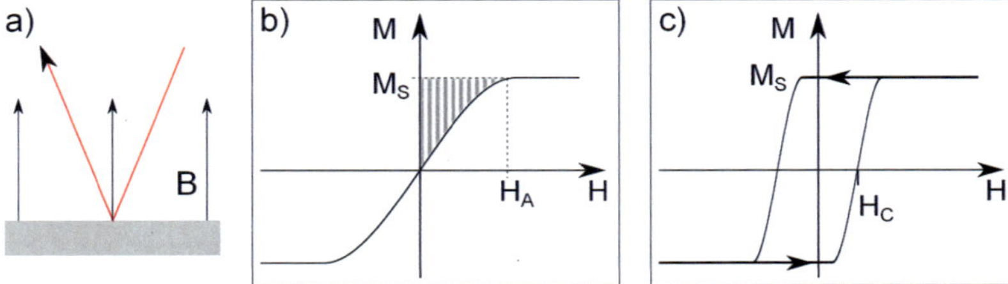

Figure 2.1: a) Polar MOKE configuration with the incident light (red) probing the out-of-plane component of the magnetization in a magnetic field perpendicular to the surface. b) Hysteresis loop of a sample with in-plane easy axis of magnetization measured with polar MOKE. The hatched area is equivalent to the anisotropy energy. c) Hysteresis loop with out-of-plane easy axis.

to the film plane. The absolute value of the magnetization, however, cannot be obtained by a MOKE measurement. Usually, the signal is measured during a sweep of the applied magnetic field and hysteresis loops are obtained (see Fig. 2.1b), c)). In the case of an in-plane easy axis of magnetization, the Kerr signal increases proportionally to the applied field until it saturates (see Fig. 2.1b)). This curve can be calibrated to absolute values of magnetization if the value of the saturation magnetization M_s is known. Then the magnetic anisotropy can be extracted as $E = \int \mu_0 H dM$ (hatched area in Fig. 2.1b), see e.g. [16]). The hysteresis loop in the case of an out-of-plane easy axis of magnetization is shown in Fig. 2.1c). From this measurement, we cannot access the magnetic anisotropy. The characteristic parameters are M_s and the coercive field H_c which may depend both on the resistance of the film against domain wall motion and magnetization reversal of individual domains.

2.1.3 Tunnel magnetoresistance

The basics of the tunnel magnetoresistance (TMR) effect shall be recalled in this part. As was first discovered by Jullière [17], the resistance of a tunnel junction with two magnetic electrodes (see upper part in Fig. 2.2) can be varied by changing the magnetization direction of one of the electrodes with respect to the other. We will consider the simple case of two identical electrodes and a low voltage applied to the junction. In a ferromagnet, the density of states shows a splitting of the energy bands for the different spin directions, $n_\uparrow$ and $n_\downarrow$, due to the exchange interaction. This may result in a finite spin polarization of the density of states at the Fermi energy:

$$P = \frac{n_\uparrow(E_F) - n_\downarrow(E_F)}{n_\uparrow(E_F) + n_\downarrow(E_F)} \neq 0. \tag{2.5}$$

Assuming a spin-conserving process, we can describe the transmission current through the tunnel junction as the sum of the two spin currents. According to Fermi's golden rule, each of them is proportional to the product of the corresponding spin densities of states of the two electrodes at the Fermi level n^i, n^f:

$$I = I_\uparrow + I_\downarrow \propto n_\uparrow^i \cdot n_\uparrow^f + n_\downarrow^i \cdot n_\downarrow^f. \tag{2.6}$$

In the case of two identical electrodes, there are only two different values for n. Then, the current I_p for the parallel alignment of the spin polarizations ($n_\uparrow^i = n_\uparrow^f, n_\downarrow^i = n_\downarrow^f$) (see Fig. 2.2a)) is always higher than the current I_{ap} for an antiparallel configuration ($n_\uparrow^i = n_\downarrow^f, n_\downarrow^i = n_\uparrow^f$) (see Fig. 2.2b)):

$$(I_p - I_{ap}) = [(n_\uparrow^i)^2 + (n_\downarrow^i)^2 - 2 \cdot n_\uparrow^i \cdot n_\downarrow^i] = (n_\uparrow^i - n_\downarrow^i)^2 \geq 0. \tag{2.7}$$

A configuration with angles $0\,^\circ < \theta < 90\,^\circ$ between the spin polarization vectors P^i, P^f was first theoretically described by Slonczewski

[18] as follows:

$$I \propto \bar{I}(1 + const. \cdot P^i P^f cos\theta).\qquad(2.8)$$

Usually, the TMR effect is studied in epitaxial tunnel junctions with electrodes of different coercive fields. This allows to switch the magnetization of only one electrode with a magnetic field. A typical experiment is sketched in Fig. 2.2b), starting with a parallel alignment of the two electrodes: when the magnetic field is increased from zero, first the electrode with low coercive field switches, which leads to an antiparallel alignment of the magnetizations (spin polarizations) and a high resistance R_{ap}. At even higher magnetic fields, also the second electrode switches, and both magnetizations (spin polarizations) are again aligned in parallel, which leads to the low-resistance state R_p. Then the TMR ratio is described as $(R_{ap}-R_p)/R_p$, where R_{ap} and R_p are the resistances for parallel and antiparallel alignment of the magnetization of the two electrodes. Note that the magnetization is given by the integrated density of states, while the spin polarization depends on the values of the spin density close to the Fermi energy. Reversal of the magnetization, however, also leads to reversal of the spin polarization.

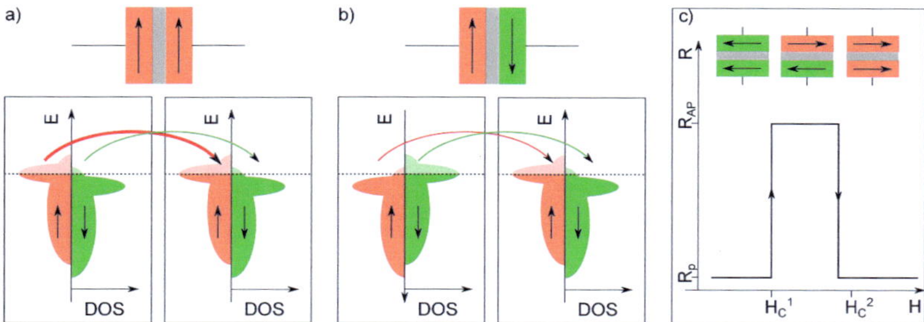

Figure 2.2: a) Spin-dependent tunneling in a TMR junction for parallel and antiparallel spin polarizations of the electrodes. The Fermi level is indicated by the dashed line. b) Resistance of a magnetic tunnel junction composed of two electrodes with coercive fields $H_c^1 < H_c^2$ as a function of the magnetic field.

2.2 Scanning tunneling microscopy

A scanning tunneling microscope consists of a metallic tip that is scanned closely above a conductive surface. A voltage applied between tip and sample leads to a tunneling current which is used to gather information about the surface. In this chapter, the history of scanning tunneling microscopy (STM), the basic theoretical description and the different operation modes used throughout this thesis will be presented.

2.2.1 History

The first basic ideas of a scanning probe instrument, called the 'Topografiner', date back to 1972 [19]. Several volts applied between tip and sample lead to a distance-dependent field emission current. Though its mediocre resolution, this instrument gave birth to a whole family of microscopes that are scanning the sample surface and probing the variable of interest pixel by pixel. This real-space technique was opposed to diffraction techniques that were dominating surface science at these times. In 1981, Binnig and Rohrer built the first working STM [20, 21] with which one year later atomically resolved images of surfaces were recorded. Only four years later, they were awarded the Nobel prize. Today, STM is widely used in surface science and plays a major role in the rising field of nanotechnology.

2.2.2 Theory of tunneling

The main variable in STM is the quantum mechanical tunneling current that flows at finite applied voltages when the tip-sample separation is inferior to about $1\,\mathrm{nm}$. In a classical description, electrons are not allowed to overcome the vacuum barrier between the two metal electrodes as long as their energy E is lower than the barrier Φ. The height of this barrier (which is approximately equal to the work function) is typically in the range of a few electron volts, which is usually higher than the energy of the electrons due to the applied voltage. Quantum mechanics tells us that the electron wave functions decay exponentially into this barrier and thus, there is a finite probability to find the electron on the other side of the barrier. Schrödinger's equation for this one-dimensional problem gives a quantitative description of the tunneling probability of an electron (described by the wave function Ψ) impinging on the barrier (see Fig. 2.3a)):

$$[-\frac{\hbar^2}{2m}\frac{\partial}{\partial z} + V(z) - E]\Psi(z) = 0, \qquad (2.9)$$

where the potential is $V = \Phi$ at the barrier and $V = 0$ elsewhere. A general solution for the three regions I, II, III, depicted in Fig. 2.3a), is given by

$$\Psi(z) = Ae^{kz} + Be^{-kz} \text{ with } k = \frac{\sqrt{2m(V-E)}}{\hbar}. \qquad (2.10)$$

As the partial solutions for the wave function and its derivatives have to match at the boundaries, the wave function Ψ and from this the transmission probability T can be calculated. In the STM configuration, the barrier width z_0 is usually large compared to the wavelength, i.e., $kz_0 >> 1$, and it is found:

$$I \propto |T^2| \propto= e^{-2kz_0}. \qquad (2.11)$$

This simple picture gives an idea of why STM is so sensitive in z-direction: assuming that the tunneling current is directly proportional to T, it changes by about one order of magnitude when z_0 changes by $0.1\,\mathrm{nm}$ (for typical values of Φ and U).

The first description of a tunneling current between two metallic electrodes μ and ν was presented by Bardeen in 1961 [22]. In this model, the Schrödinger equation of each electrode is solved independently, and the overlap of the two wave functions on a virtual surface within the barrier is considered.

For an elastic tunneling process, the energies of the initial state E_μ and of the final state E_ν must be the same. The contribution of the states μ and ν to the tunneling current depends on the product of the overlap of the wave functions and the population of each state (which is given by the Fermi function $f(E)$). At an applied voltage U, the tunneling current between μ and ν is then given by the sum over all states:

$$I = \frac{2\pi e}{\hbar} \sum_{\mu,\nu} [f(E_\mu)[1-f(E_\nu+Ue)] - f(E_\nu+Ue)[1-f(E_\mu)]] \quad (2.12)$$

$$\times \, |M_{\mu\nu}|^2 \delta(E_\mu - E_\nu), \quad (2.13)$$

where $E_{\mu,\nu}$ is the Fermi energy of the corresponding electrode. The tunneling matrix element M in its general form is given by:

$$M_{\mu\nu} = \frac{\hbar^2}{2m} \int d\vec{S} \cdot (\Psi_\mu^* \vec{\nabla} \Psi_\nu - \Psi_\nu \vec{\nabla} \Psi_\mu^*). \quad (2.14)$$

It can be calculated under certain approximations. The first approach for the STM geometry was given by Tersoff and Hamann [23]. They modeled the tip apex with a spherical potential barrier and s-type wave functions decaying exponentially into the vacuum. This allows to express the tunneling matrix element as a function of the local density of states (LDOS) of the sample. In the approximation of low applied voltages U and low temperatures (the latter simplifies $f(E)$ to a step function), the tunneling current is proportional to the integrated density of states of sample and tip:

$$I \propto \int_0^{eU} \rho_{sample}(r_0, E_{F,S} + E - Ue)\rho_{tip}(E_{F,T} + E)dE, \quad (2.15)$$

with $\rho_{sample}(r_0, E_{F,S} + E)$ being the density of states of the sample at the tip position (see Fig. 2.3b), c)). The density of states of the tip ρ_{tip} is often assumed to be constant within a small voltage range. Then the first derivative of the tunneling current with respect to the applied voltage is proportional to the LDOS of the sample at $z = r_0$:

$$\frac{\partial I}{\partial U} \propto \rho_{tip}(E_F)\rho_{sample}(r_0, E_F + eU). \tag{2.16}$$

For higher applied voltages, a further simplification allows to express $\rho_{sample}(r_0, E_{F+eU})$ as a product of the LDOS at the surface and the transmission coefficient T. Then the differential conductance can be written as

$$\frac{\partial I}{\partial U} \propto \rho_{tip}(E_F)\rho_{sample}(E_F + eU) \cdot T. \tag{2.17}$$

In the WKB approximation [24], T is

$$T \approx exp(-2z\sqrt{\frac{2m}{\hbar}(\frac{\Phi_{sample} + \Phi_{tip}}{2} + \frac{eU}{2} - E))}. \tag{2.18}$$

So, the normalized differential conductance $\frac{\partial I}{\partial U}/T$ gives an estimate of the LDOS of the sample surface.

2.2.3 Operation modes

The fact that the measured tunneling current includes not only information about the geometry of the sample but also about its electronic structure is both a blessing and a curse. Although STM is considered to be a straightforward method for surface science, it can be difficult to disentangle the different contributions to the tunneling current. Therefore, different measuring modes are necessary to obtain valuable information.

The used control variables for STM are the bias voltage U, the lateral position x, y and either the tip-sample separation z or the

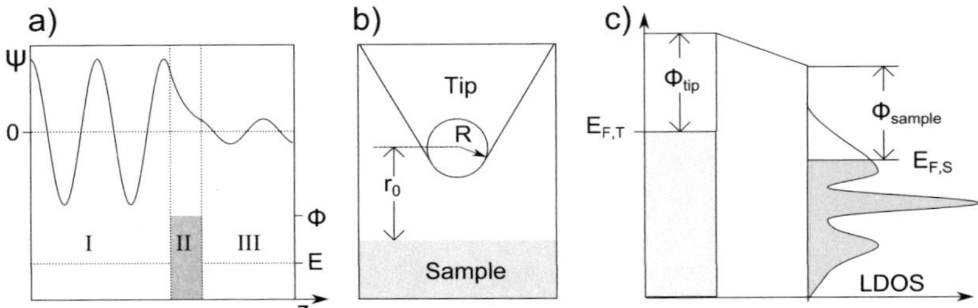

Figure 2.3: a) Real part of the electron wave function and energy for tunneling through a one-dimensional barrier. b) Configuration of tip and sample in the Tersoff-Hamann model. c) Tunneling in the STM setup with a constant tip LDOS: the barrier is formed by the work functions of tip and sample.

setpoint for the tunneling current I. In the following, the measuring modes for the dependent variables will be summarized. The time dependence will be explicitly mentioned in case it has been used for the measurements presented later in this thesis.

Tunneling current I

The tunneling current is the basic variable in STM and is directly measured using an I-V converter. In order to record time-dependent phenomena of the sample, the tunneling current is recorded as a function of time at fixed tip position: I vs. t with $U, z, x, y = $ const. Insight into the electronic structure can be obtained by recording the current as a function of the applied bias voltage: I vs. U, t with $z, x, y = $ const. Measuring the exponential distance dependence of the current allows to deduce the local work function: I vs. z with $U, x, y = $ const.

Tip-sample separation z

Measuring z is only meaningful when we use the feedback loop which controls the tip-sample separation in order to keep the tunneling current constant. Mapping z, or more precisely the voltage applied to the corresponding piezo, is the basic operation mode for STM, the constant current mode (z vs. x, y with $I, U = $ const.). The resulting image, called topography, displays the contour of a constant sample density of states which is mainly given by the geometry of the sample (the density of states decays exponentially with distance). Although one might argue that the 'real' surface structure is given by the position of the atomic cores, many properties are dominated by the electronic structure which is imaged by STM. Measuring z at laterally fixed tip position as a function of time (z vs. t with $x, y, U, I = $ const.) gives similar information as I vs. t with usually better signal-to-noise ratio but lower time resolution. This can be combined with a voltage variation (z vs. t, U with $x, y, I = $ const.). In order to detect spatially different time dependencies, this mode is combined with an x-scan, called linescan: z vs. t, x, U with I, y $= $ const.

Differential conductance dI/dU

As explained above (see paragraph 2.2.2), the voltage derivative of the tunneling current dI/dU is approximately proportional to the local density of states (LDOS). This differential conductance is acquired by using lock-in technique. When a sinusoidal modulation is superimposed to the bias voltage, the corresponding Fourier component of the tunneling current is proportional to differential conductance dI/dU. It is often recorded as a function of the applied bias voltage at fixed tip position, usually referred to as scanning tunneling spectroscopy (STS). In this work, it is also used in order to observe a possible time dependence of the sample structure because of the high signal-to-noise ratio of the lock-in signal: dI/dU vs. U, t with $z, x, y = $ const. The dI/dU signal can also be collected in the mapping mode simultaneously with the topography, which results in a map of the density of states at the energy given by the bias voltage: dI/dU vs. x, y with $U, z = $ const. The last two techniques will also be combined (for spatially resolved electric-field-dependent switching dynamics): dI/dU vs. U, x, y, t with $z = $ const.

Differential work function dI/dz

Mapping of the local apparent barrier height (roughly: the work function) allows to record high-resolution images, interpreted similarly as topographic images: dI/dz vs. x, y with $U, z = $ const.

2.2.4 Electric field in STM

Usually, the electric field in the STM junction will be estimated as the applied bias voltage U divided by the tip-sample separation z. In Fig. 2.3c), it can be seen that the potential difference, that actually determines the value of the electric field in between tip and sample, is given by $\Delta V = U + (\Phi_{sample} - \Phi_{tip})$.

The tip-sample separation z is not directly accessible in STM. By using a simple approximation, it can be derived from tunneling current I, bias voltage U and the local work function Φ:

$$I = \frac{1}{R_0} \cdot U \cdot e^{-2z\sqrt{2m\Phi}/\hbar} \Leftrightarrow z = \frac{-\hbar}{\sqrt{2m \cdot \Phi}} \cdot ln(\frac{I}{U} \cdot R_0),$$

$$(2.19)$$

where R_0 is the resistance in contact at $z = 0$. It can be approximated by the resistance of a single atomic point contact to $R_0 = 12900\,\Omega$ [25]. The work function was measured as shown in Fig. 2.4a). As STM measurements are usually carried out in the constant current mode, the distance z changes when the applied

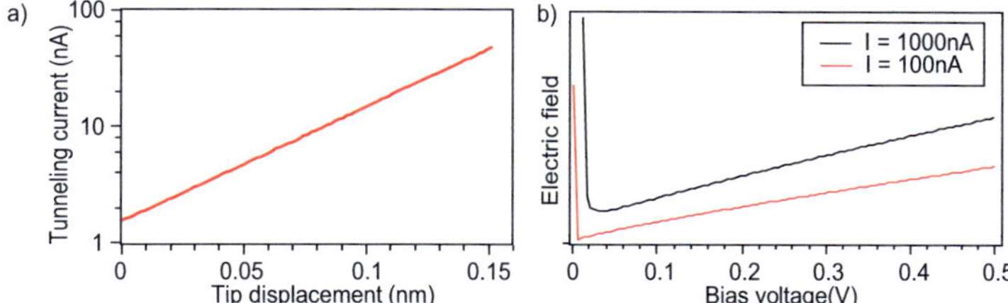

Figure 2.4: a) Measured increase of I as a function of z in order to determine the work function. b) Electric field as a function of applied voltage in the constant current mode for two different tunneling currents following the approximation given in the text.

bias voltage U is changed. However, the above-given approximation shows that for all combinations of U and I which are typically accessible with STM the electric field scales monotonically with the voltage (see Fig. 2.4b)). As a rule of thumb, $E \propto U$ is also correct in the constant current mode and therefore a variation of the electric field is usually achieved by variation of the bias voltage. Besides the possibility of a linear voltage variation with time, our STM software supplies a special feature to apply voltage pulses with defined length and amplitude during which the feedback loop is switched off.

2.3 Magnetoelectric coupling

Growing interest in magnetoelectric coupling (MEC) is driven by its potential for future memory technologies. Conventional magnetic data storage devices require magnetic fields to write the data: the magnetization direction of each bit is controlled by the high stray field of the electromagnet in the write head. Using electric fields to control the magnetization via MEC could open up a new pathway to high-density low-power data storage devices. In the following, a simple introduction to MEC and the progress in this field are presented. A brief review on previous work on MEC in metallic thin films sketches the context of this thesis.

2.3.1 Definition of MEC

MEC describes the bidirectional influence of magnetic and electronic properties. So far, this phenomenon has mainly been observed in insulating materials such as complex multiferroic oxides. Multiferroic materials spontaneously display more than one of the following ferroic orders in a single phase: ferromagnetism, ferroelectricity and ferroelasticity. These might couple to one another directly, indirectly or by their corresponding fields. These possible interactions are shown in a schematic view in Fig. 2.5a). While many materials are known that show magnetoelastic or piezoelectric (electroelastic) behavior, only few single-phase magnetoelectric compounds exist. Although strong MEC is expected for materials that are ferromagnetic and ferroelectric at the same time, ferroism is not indispensable for MEC (see Fig. 2.5b)).

2.3.2 Basics

In a general way, following Landau theory, the free energy F of a single phase can be written in terms of the applied magnetic field

H, the electric field E and the stress σ:

$$F(E,H,\sigma) = F_0 - P^s E - \mu_0 M^s H - \epsilon_S \sigma - \frac{1}{2}\epsilon_0 \chi_E E^2 - \frac{1}{2}\mu_0 \chi_M H^2 \tag{2.20}$$

$$- \frac{1}{2}s\sigma^2 - \alpha E H - dE\sigma - qH\sigma - ..., \tag{2.21}$$

where F_0 is the energy in absence of any field, P^s, M^s and ϵ_S are the spontaneous polarizations, ϵ_0, μ_0 and s are the susceptibility constants.

The second, third and fourth term denote the interaction of the electric field with the spontaneous polarization, the magnetic field with the spontaneous magnetization and the stress with the elastic deformation, respectively. The electrostatic energy $\frac{1}{2}\epsilon_0 \chi_E E^2$ scales with the electric susceptibility χ_E, the magnetostatic energy with the magnetic susceptibility χ_M and the elastic energy $\frac{1}{2}s\sigma^2$ with the elastic tensor s. The following terms denoted here are the lowest-order contributions given by a cross correlation of electric field, magnetic field and stress which are given by the corresponding coupling constants: MEC (α), piezoelectric (d) and piezomagnetic (q) coupling. Usually, higher-order magnetoelectric effects are omitted, and the prefactor α which defines the linear coupling between magnetic field and electric field is called magnetoelectric coefficient and is used to compare the magnitude of MEC in different systems. Then both the electric polarization induced by a magnetic field and the magnetization induced by an electric field scale with α:

$$P(E,H,\sigma) = -\frac{\partial F}{\partial E} = P^s + \epsilon_0 \chi_E E + \alpha H + d\sigma + ..., \tag{2.22}$$

$$M(E,H,\sigma) = -\frac{\partial F}{\partial H} = M^s + \mu_0 \chi_M H + \alpha E + q\sigma..., \tag{2.23}$$

$$\epsilon(E,H,\sigma) = \frac{\partial F}{\partial \sigma} = \epsilon_S + s\sigma + dE + qH + ... \tag{2.24}$$

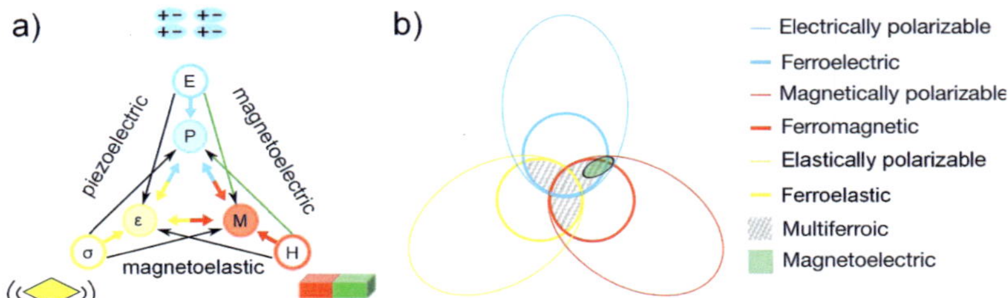

Figure 2.5: a) Interaction triangle of electric field E, stress σ, magnetic field H and the corresponding order parameters electric polarization P, elastic deformation ϵ and magnetization M (after [6]). Particularly interesting is the possible influence of an applied electric field on the magnetization (shown as green arrows). b) Classification of polarizable materials: in multiferroic materials (hatched area), at least two ferroic orders (thick colored circles) coexist; MEC could appear in any material which is both magnetically (thin red circle) and electrically polarizable (blue thin circle), and it is not necessarily limited to ferroic materials (after [26]).

The prerequisites for MEC can be extracted from symmetry arguments when the different ferroic properties are considered: while the electric polarization is invariant under time inversion but reverses under space inversion, the magnetization is invariant under space inversion but changes sign under time inversion. Thus, MEC (in multiferroics) invariably requires a system in which both space and time inversion symmetry are broken.

2.3.3 Development of MEC in insulators

The first experiments date back to 1888 when Röntgen observed a finite magnetization in a dielectric that changed in the presence of an electric field [27]. Although the necessity of broken symmetry was already pointed out by P. Curie in 1894 [28], it took until the 1960s that Soviet scientists found the first material exhibiting static magnetoelectric coupling [29, 30]. Dzyaloshinskii found that piezoelectric Cr_2O_3 displays long-range magnetic order and thus breaks time reversal symmetry [31]. To date, this phenomenon has been observed in many different compounds that are mainly insulating materials such as complex oxides [32]. They can be classified into two groups: single-phase magnetoelectric materials combine both electric and magnetic dipole moments in the same phase. The magnetoelectric effect in these materials is very weak, roughly 10^{-5} of the magnetization is affected by the electric field [6]. Furthermore, most of these materials display very low ordering temperatures. In typical ferroelectric materials like the perovskites, the mechanisms leading to ferroelectricity and ferromagnetism even seem to exclude each other: a stable electric polarization arises only in the case of empty d-shells of the transition-metal ions, while partially filled d-shells are responsible for the ferromagnetism [33]. A second group are composite materials in which ferroelectric and ferromagnetic phases are brought into close contact so that electric and magnetic dipoles couple via the interface, driven by elastic [5, 34] or electronic effects [8, 7]. Especially in laminated multilayer compounds, larger effects are found. In both approaches, the electric polarization (the position of the ions in the ferroelectric material with respect to one another) can be switched by an applied electric field. This leads to microscopic changes in the interatomic distances, which in turn changes the magnetic properties.

2.3.4 MEC in metals: context of this work

Bulk metallic systems do not exhibit MEC because applied electric fields are screened by a surface charge. In a metal, unlike in semiconductors for example [35], the high charge carrier density restricts the extension of this screening charge towards the bulk to the first few atomic layers. But in the very surface, the potential impact on the crystallographic and electronic properties is high.

Given a high surface-to-volume ratio, this surface effect might substantially influence the properties of the whole system. This manner of MEC in metallic thin films was first predicted by Blügel in 2000 [10]. In 2003, Weissmüller et al. found a reversible surface-charge-induced strain in non-magnetic platinum [9]. The first experimental evidence for an electric-field-induced change of the magnetic properties of a metallic thin film was given by Weisheit et al. [11]. They observed a variation of a few percent in the coercive field of FePt and FePd films under the application of electric fields (see Fig. 2.6a)). It was claimed that the electric field changed the number of unpaired d-electrons in the surface and thus affected the magnetic properties. The next step towards a possible application of MEC in high-efficiency memory devices was made in the group of Y. Suzuki: a change of the magnetic anisotropy by 40 % induced by an electric field in a Fe/MgO junction was observed in 2009 by Maruyama et al. [12] (see Fig. 2.6b)). The presence of a constant electric field may influence the magnetic order, but it does not break time reversal symmetry and thus cannot be used to switch between two bistable magnetization directions of a ferromagnet. Two years later, in the same group, Shiota et al. [13] showed that in a TMR junction the magnetization can be switched between two orientations by nanosecond voltage pulses. Using asymmetrically shaped voltage pulses, they could break time reversal symmetry and induce a coherent rotation of the magnetization.

As was shown in 2.2, an STM is an ideal tool with which to investigate MEC because it has the capability to image surface structures with atomic resolution and it provides the possibility to use the high electric field underneath the STM tip (see Fig. 2.6c)). Making use of these advantages, in this work we bring the first proof of principle for bistable switching of the magnetic order in a metal thin film using MEC. In contrast to any other technique, STM allows us to only locally induce very high electric fields and thus to trigger the magnetism on the nanometer scale. The reverse of these advantages is the bad time resolution of STM. Regarding symmetry arguments, any ferromagnetic surface (or interface) may display MEC: space inversion symmetry is always broken by the surface and time inversion symmetry by the ferromagnetic order. In order to obtain strong effects, it is favorable to work on systems at a magnetic phase transition which then could be triggered by the applied electric field. A phase transition of first order would provide the possibility of toggle switching between two bistable states. In the STM studies on 2 ML Fe/Cu(111) and 1 ML Fe/Ni(111) presented in this work, the electric field indirectly acts on the magnetic order via strain. The electric field induces a phase transition of first order in the crystallographic structure that is clearly identified with atomic-resolution STM. This change of the crystallographic structure in turn changes the magnetic order in 2 ML Fe/Cu(111) from FM to AFM (see chapter 4) and the magnetic properties of 1 ML Fe/Ni(111) (see chapter 5). Furthermore, MEC in 10 ML Ni/Cu(100) is studied in solid-state insulator junctions: here, the observed change of coercive field and saturation magnetization will be attributed to the direct influence of the electric field on the local electronic properties of the surface (see chapter 6).

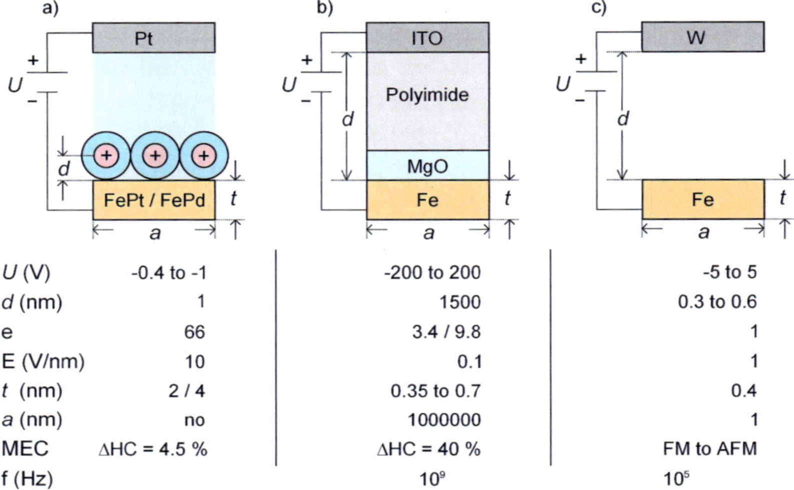

	a)	b)	c)
U (V)	-0.4 to -1	-200 to 200	-5 to 5
d (nm)	1	1500	0.3 to 0.6
e	66	3.4 / 9.8	1
E (V/nm)	10	0.1	1
t (nm)	2 / 4	0.35 to 0.7	0.4
a (nm)	no	1000000	1
MEC	ΔHC = 4.5 %	ΔHC = 40 %	FM to AFM
f (Hz)		10^9	10^5

Figure 2.6: Three important experiments on MEC in metallic thin films: a) Change of coercive field in FePd and FePt films by an electric field applied through a double layer in an electrolyte, Weisheit, Givord et al. (2007), b) Large electric-field-induced change of anisotropy in a MgO/Fe junction, Maruyama & Suzuki et al (2009), c) MEC in STM presented in our work (published in 2010). Comparison of characteristic parameters: applied voltage U, thickness of insulator d, dielectric constant ϵ, electric field E, thickness of the magnetic film t, size of the influenced area a, MEC (change of coercive field in a) and b), transition between FM and AFM in c)) and maximum switching frequency f.

3 Experimental setup

3.1 STM and UHV preparation chamber

Measurements on MEC in Fe/Cu(111) and Fe/Ni(111) were carried out with a low-temperature STM (see Fig. 3.1a), b)) in the group of Prof. Wulfhekel. The employed preparation methods and analysis tools as well as the construction of the STM will be briefly presented in this paragraph.

Vacuum

STM is sensitive to the topmost atomic layer of a surface, therefore the basic task in the preparation of an STM sample is to obtain an atomically clean surface. This is why the whole experimental procedure is carried out in ultra-high vacuum (UHV) at pressures of about $1 \cdot 10^{-10}$ mbar. This corresponds, for example, to a contamination rate of one molecule per nm^2 in one hour at room temperature. Inside the cryostat, where the STM is located, the pressure is even significantly lower. This vacuum level is achieved by using a standard two-stage pumping system: a rotary pump reduces the pressure to about $1 \cdot 10^{-3}$ mbar, and turbo-molecular pumps further decrease the pressure to the UHV regime. An initial out-gassing by heating the UHV chamber is necessary to remove contaminants from the walls and is repeated each time the chamber is exposed to air. Once UHV conditions are achieved, an ion pump suffices to maintain the vacuum during the measurement. As the mean free path of molecules in this pressure regime is many orders of magnitude higher than the diameter of the chamber, desorption from the walls is crucial. A temporal further reduction of the pressure can be achieved by sublimation of titanium (which incorporates impinging gas molecules) to the inner chamber walls and/or cooling them with liquid nitrogen in order to reduce desorption.

Substrate preparation

The cleaning process of the substrates follows a standard procedure of repeated cycles of bombardment with Ar^+ ions of 1.5 keV, called sputtering (see Fig. 3.1), and heating to elevated temperatures, called annealing. While the sputtering is intended to remove adsorbates from the surface, annealing heals defects by mobilizing the surface atoms. Inside the UHV system, the samples are carried on sample holders which can be transported by a system of manipulators and wobble sticks. For the STM measurements, two different substrates were used: a Cu(111) single crystal and a Ni(111) single crystal. The crystals have a hat shape and are mounted on a sample holder in such a way that they can be annealed by direct electron bombardment (see Fig. 3.2a)). This avoids undesirable heating of the sample holder and its surroundings (thus desorption) during the annealing process and reduces the time for cool-down, which is advantageous when absolute cleanness of the surface is needed. The quality of the surface can be studied by low-energy electron diffraction (LEED) and Auger electron spectroscopy (AES) (see Fig. 3.1a)), but the ultimate check can only be done with STM.

Deposition of Fe

Once the sample surface is cleaned and cooled down to room temperature, Fe is deposited by molecular beam epitaxy (MBE). A pure (99.99 + %) Fe rod is heated by bombardment of electrostatically accelerated electrons thermally emitted from a filament in

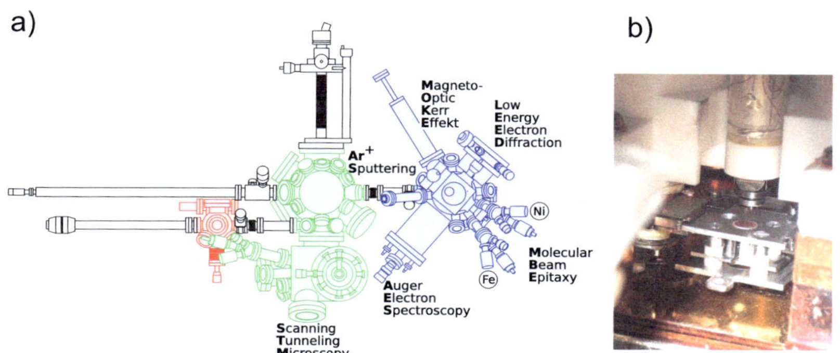

Figure 3.1: a) Vacuum chamber at KIT: preparation chamber with basic surface preparation and analysis tools (blue), STM chamber with He cryostat (green) and two-chamber load lock (red); from [36]. b) View into the STM.

front of the rod (see Fig. 3.2b)). The evaporated Fe is partially ionized and thus a small current proportional to the evaporation rate of Fe can be detected at the aperture. Usually, sub-monolayer amounts are deposited so that the calibration of the evaporation rate is done by STM. Typical heating parameters of an emission current $I_{em} = 10\,\mathrm{mA}$ and an acceleration voltage of $HV = 900\,\mathrm{V}$ result in a flux of ionized Fe of $I_{flux} = 30\,\mathrm{nA}$ and a deposition rate of $90\,\mathrm{s/ML}$.

Cryostat

All STM measurements described in this work were carried out with an STM situated inside the shieldings of a He^4 bath cryostat. The He^4 cryostat is surrounded by an additional nitrogen cryostat which increases the measurement time at low temperatures to about $40\,\mathrm{h}$. Low temperatures are beneficial for STM measurements in several ways: firstly, the pressure, especially the partial pressure of hydrogen, is decreased by the cold surfaces of the cryostat which traps the molecules from the gas phase. Secondly, the atomic arrangement of the surface is stabilized because thermal diffusion is strongly suppressed. Finally, the highest possible energy resolution in the spectroscopic mode is increased.

Damping

Besides the UHV conditions and the low temperature, a low mechanical noise level is important for STM. Therefore the whole setup is carried by pneumatic insulators with laminar-flow damping in order to separate out mechanical noise. External vertical and horizontal vibrations are suppressed by more than one order of magnitude for frequencies above 10 Hz. A second mechanical filter is realized by four springs suspending the STM body from the cryostat. Magnetic fields of about 50 mT produced by superconducting coils provide additional efficacious damping by inducing eddy currents in the copper parts of the STM.

STM

The construction frame of the home-built STM itself is made of copper and holds a sample stage which can be moved by piezo motors in x and y direction. The tip is exchangeable in situ and mounted on a tube piezo for the scanning fine motion. The tube piezo can be moved in z direction for the approach of the tip into the tunneling regime by another set of piezo motors. No magnetic materials are used in order to avoid any stray fields that could influence the magnetism of the sample. A view into the STM is shown in Fig 3.1b).

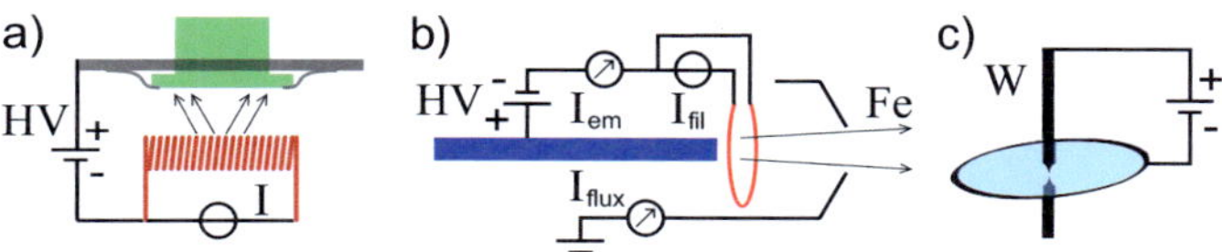

Figure 3.2: a) Sample holder used for heating hat-shaped single crystals (green) by electron bombardment. b) Evaporation of Fe from a rod (blue) by electron bombardment, for details see text. c) Electrochemical etching of a sharp tip from a tungsten wire immersed in a soap film of NaOH.

STM tip

The tips used for the STM measurements are electrochemically etched from $0.3\,\mathrm{mm}$ tungsten wire by using NaOH (see Fig. 3.2c)). After the tip has been inserted into the UHV chamber via the load lock (see Fig. 3.1), it is cleaned by sputtering with Ar^+ ions of $3\,\mathrm{keV}$ and subsequent melting of the tip apex by electron bombardment, which also promotes a more stable configuration of the tip apex.

STM electronics

The tunneling current is measured by a commercial *I-V* converter by Femto using amplifications of 10^5-10^9 V/A and a bandwidth of 1-$200\,\mathrm{kHz}$. For the measurements of the derivatives of the tunneling current, an analogue lock-in amplifier from Physical Instruments is used. The tip motion, the feed-back loop parameters and the applied bias voltage are controlled by analogue electronics from RHK Technology. The measurement software which is used to acquire images and spectra is also supplied by RHK Technology.

3.2 Setup for MEC measurements with electric fields applied through solid-state insulators

The experiments on MEC in MgO/Ni/Cu(100) multilayer systems were done during my stay in Prof. Suzuki's group, Osaka University. The equipment needed to prepare these samples and the measurement tools for MEC studies at solid-state insulator interfaces will be discussed shortly.

MBE

For the deposition of highly homogeneous metal films on $4\,cm^2$ samples, a special MBE setup operating at UHV (base pressure of $5 \cdot 10^{-10}\,mbar$) was used. Polished MgO wafers introduced via a load lock were used as substrates. They were heated to $500\,°C$ by radiation from a filament close to the sample in order to clean the surface. Three evaporators each containing five exchangeable crucibles allow the growth of a large variety of metallic thin films including alloys made by co-deposition and insulating barriers like MgO. An electron beam with energies of up to $10\,keV$ and $300\,mA$ maximum emission current heats the material to be evaporated which is held by a water-cooled molybdenum crucible. With the help of magnetic and electric fields, the electron beam can be focused and swept across a defined area on the crucible. A manually adjustable linear shutter located just in front of the sample allows the growth of wedge-shaped films with a precise thickness gradient. The thickness of the metal film is monitored by a calibrated quartz microbalance. Additionally, the film growth can be followed by reflection high-energy electron diffraction (RHEED). With this setup, multiple layers with thicknesses of up to several tens of nanometers can be deposited with high accuracy within reasonable time.

Microfabrication

Electron beam lithography (EBL) was used for microfabricating the magnetic TMR junctions. As lift-off technique with a specially designed resist and Ar-milling in combination with a mass spectrometer is used, structures with lateral dimensions down to about 200 nm can be prepared. The computer-controlled EBL allows the preparation of several hundred junctions in one step (see Fig. 3.3a)). These facilities are located in a clean room which is also equipped with a further chamber that was used for deposition of SiO_2.

MOKE and TMR measurement setup

Both MOKE and TMR experiments were carried out at room temperature in air. The MOKE measurements were performed in a polar configuration which probes the out-of-plane component of the magnetization. A simple setup of conventional Cu-coils and pole shoes allows to apply magnetic fields perpendicularly to the film plane of up to 16 000 Oe (see Fig. 3.3b)). A polarized and focused laser beam passes through holes in the pole shoe, is reflected from the sample and after it has passed through a second polarizer the intensity is measured with a photo diode. The sample can be moved by stepping motors in the plane perpendicular to the magnetic field in order to address different thicknesses of a wedge-shaped sample. The electric field is applied by contacting the sample with a conventional Keithley power source. A LabView software is used to acquire the measured hysteresis loops and to ramp the applied voltage automatically. The measurement setup for the MTJs consists of a pair of magnetic coils and pole shoes that produce a magnetic field in the sample plane of up to 3000 Oe (see Fig. 3.3c)) and a special three-fingered prober to contact the submillimeter-sized junctions. It is approached to each MTJ by using millimeter screws in an optical microscope. The voltage is

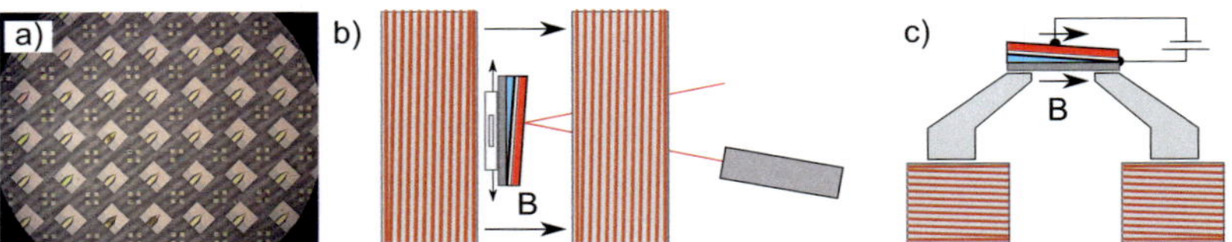

Figure 3.3: a) Array of identical tunnel junctions prepared by EBL (field of view $4 \times 3 \, \mathrm{mm}^2$). b) Measurement setup for polar MOKE with a magnetic field applied perpendicularly to the sample plane. c) Measurement setup as used for TMR measurements with the magnetic field applied in-plane.

applied using a Keithley power source which is also used to measure the resulting tunneling current. A LabView program is used to repeat the magnetic field sweeps and to set the voltage applied through the tunnel junction.

4 Experiments on 2 ML Fe/Cu(111)

In this part, the experimental results and corresponding theoretical calculations on the system of 2 ML Fe/Cu(111) will be discussed. The basic sample preparation procedure and the crystallographic and magnetic structure of this system are presented. Then, experimental evidence for MEC is given, and the dynamics of the switching process are discussed in more detail. The magnetic identification of the phases by differential conductance spectra as well as the discovery of the electric-field-induced phase transition are results which have already been discussed in my diploma thesis [37]. The experimental finding of MEC and the theoretical simulations (Fig. 4.1 to 4.7) are published results [38, 39, 40]. Fe is known for its large number of different phases. At room temperature and ambient pressure, bulk Fe is stable in the body-centered cubic (bcc) phase. At high temperatures of $1184\,\mathrm{K}$, a phase transition to the face-centered cubic (fcc) γ-phase occurs, which is important e.g. in the processes involved in steel hardening. For thin films, the phase diagram is even more complex. Fe films on Cu single crystals have been extensively studied for a long time due to the fact that in these thin films the fcc phase of Fe can be stabilized at room temperature. A structural transition from fcc to bcc with increasing thickness is reported for various orientations of the substrate surface [41, 42]. As it is known that the magnetic structure of Fe delicately depends on the volume of the crystallographic unit cell [14], also a large variety of magnetic orders is expected for Fe thin films. In the following, it will be shown that these properties make 2 ML films of Fe/Cu(111) a model system for MEC at metal surfaces.

4.1 Preparation of 2 ML Fe islands on Cu(111)

The Cu(111) surface was cleaned by repeated cycles of sputtering and annealing up to 500 °C (see chapter 3). At room temperature, a sub-monolayer amount of Fe was deposited on the atomically clean Cu(111) by MBE as described in paragraph 3.1. As the Fe surface is prone to contamination, e.g. by hydrogen, the sample is transferred to the STM chamber directly after deposition. A typical deposition of an amount equivalent to 0.8 ML Fe/Cu(111) leads to the formation of 10 to 20 nm large triangular bilayer islands (see Fig. 4.1a)) and a negligible amount of 3 ML islands. The average size of the islands can be reduced to a few nanometers when the sample is cooled to about 5 °C before deposition. The triangular shape originates from the different diffusion coefficients along the different $\langle 1\,1\,0 \rangle$ step edges of the island and was first found in the similar system of Co/Pd(111) [43].

4.2 Crystallographic structure of 2 ML Fe/Cu(111)

In an STM topography (see Fig. 4.1a)), two phases of different height can be identified on the Fe bilayer islands on Cu(111). The crystallographic structure of these two phases was first revealed and discussed in all detail in the work of Biedermann et al. [42]. They found a coexistence of an fcc domain in the center and a bcc-like phase on the rims of the islands. In our atomically resolved STM images, we identified the same crystallographic phases: in the center of the island, in a typically hexagonal domain, the atoms show a perfect hexagonal order with the lattice directions and distances of the fcc Cu substrate (see Fig. 4.1b)). At the rims, however, a small misalignment of the lattice directions with respect to the hexagonal substrate indicates a strained bcc(1 1 0) lattice. This can be understood as follows: there are three main transformation paths that describe the diffusionless, also called martensitic, transformation from an fcc to a bcc lattice. The famous Bain path [44] is characterized by the orientational relation between the initial fcc and the resulting bcc phase $(1\,0\,0)_{fcc}||(1\,0\,0)_{bcc}$. But in our case of a $(1\,1\,1)$ surface $(1\,1\,1)_{fcc}||(1\,1\,0)_{bcc}$ must be satisfied. Additionally, we observe $[1\,\bar{1}\,0]_{fcc}||[1\,\bar{1}\,\bar{1}]_{bcc}$. This is the fingerprint of the transformation path named after Kurdjumov and Sachs [45, 46]. For the case of a $(1\,1\,1)$ surface, it is convenient to split up the transformation into two deformations: the first one is an out-of-plane shearing of $19.28\,°$ which can be seen at the surface as a displacement of the toplayer atoms from threefold hollow sites to twofold bridge-positions (see Fig. 4.1c), d)). The projection on the $(1\,1\,1)$ plane is a displacement by $73.6\,\mathrm{pm}$ along the $[\bar{1}\,2\,\bar{1}]$ direction. This is the dominant deformation and will be used in the calculations to trace the fcc-to-bcc transformation. The second step is an in-plane expansion which brings the surface atomic density to that of the bcc lattice. In our case, this expansion is partially hindered by the Cu substrate so that this second step results in a strained bcc: an in-plane shearing parallel to $[1\,\bar{1}\,0]_{fcc}$ is followed by an expansion in the $[\bar{1}\,\bar{1}\,2]_{fcc}$ direction by $4.4\,\%$ and leads to a bcc surface cell with an aspect ratio of 1.50 instead of 1.42 as for ideal bcc (see Fig. 4.1e)). This leads to the misalignments between the $[0\,\bar{1}\,1]_{fcc}$ and the $[1\,\bar{1}\,1]_{bcc}$ directions of $7.2\,°$ and between the $[\bar{1}\,0\,1]_{fcc}$ and the $[0\,0\,1]_{bcc}$ directions of $3.6\,°$ that were extracted from the atomically resolved STM. As can be seen in Fig. 4.1b), the atomic positions of this model nicely fit our STM measurement.

The coexistence of the two phases arises from a competition between the energy gain in the transformation from fcc to bcc and the energy loss due to strain in the bcc domain [42]. The strain

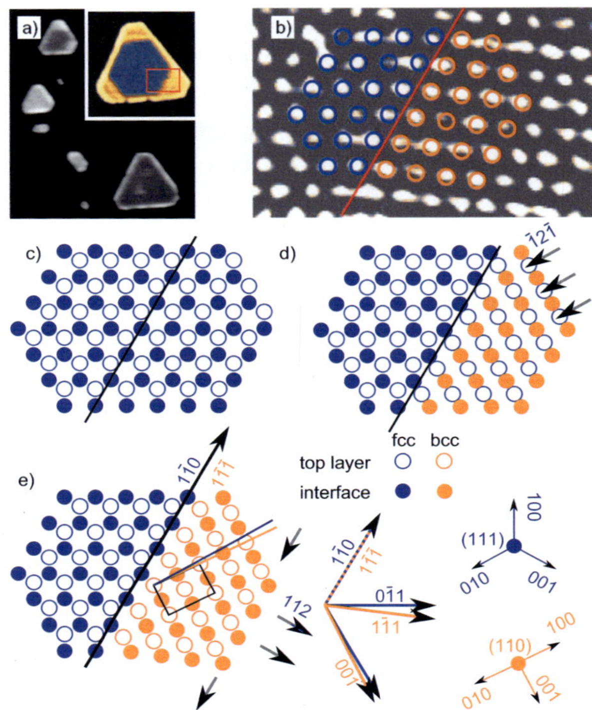

Figure 4.1: a) $77 \times 60\,\text{nm}$ STM topography of bilayer Fe islands (gray) on a Cu(111) substrate (black). Two different phases can be distinguished: a darker phase in the center and a brighter one on the rims. Small islands are fully in the bright phase, while large islands are dominantly dark. The inset shows an enlarged view with the two phases in blue and in orange. b) dI/dz map of the area marked in a) with a red rectangle showing the arrangement of the Fe atoms. The position of the domain boundary is indicated in red. The open circles indicate the atomic positions of an fcc phase in the center (blue) and a strained bcc phase (orange) as derived from the following model. c) to e) Incomplete Kurdjumov-Sachs transformation of the right part from fcc $(1\,1\,1)$ (blue) to bcc $(1\,1\,0)$ (orange). c) Ideal fcc lattice of two layers. Open circles represent the positions of the top-layer atoms. d) Out-of-plane shear that results in a displacement of the top layer from threefold hollow sites to twofold bridge positions along the $[\bar{1}\,2\,\bar{1}]_{fcc}$ direction. e) In-plane shear along the $[1\,\bar{1}\,0]_{fcc}$ direction and expansion along the $[\bar{1}\,\bar{1}\,2]_{fcc}$ direction.

of the bcc domain increases with increasing size: stacking of Fe atoms on top of the atoms in the layer underneath is energetically unfavorable and hence avoided. Therefore, a larger bcc domain can only be realized by a smaller misalignment angle with respect to the fcc lattice, thus a more strongly strained bcc. This is why the width of the bcc domains is limited to a few nanometers, and in islands of typically 10 to 20 nm only the rims of the islands crystallize in the bcc phase, while small islands can be fully bcc. The interlayer distances of the two Fe layers are not accessible with STM as the difference in the LDOS of Fe and Cu always contributes to the measured step height. At most voltages, the bcc state appears to be higher, but this contrast is inverted for small negative voltages due to a peak in the density of states of the fcc lattice (see. Fig. 4.4). In a comparison of measured differential conductance spectra with *ab-initio* calculations (see Fig. 4.4), the interlayer distance of 200 pm for the fcc phase and of 212 pm for the bcc phase (compared to 208 pm for Cu) fit the experimental data best. This finding is in agreement with the lateral strain that can be directly obtained from the atomically resolved STM (see Fig. 4.1). The lattice of the fcc domain is expanded laterally and compressed vertically. The lattice of the bcc domain is mainly compressed laterally and expanded vertically (see Fig. 4.2). The atomic volume is slighty reduced in the case of the fcc lattice, but strongly increased for the bcc lattice.

Biedermann et al. [42] found a strong influence of hydrogen adsorption on the balance of the two crystallographic structures of the Fe islands. To make sure that our experiments were carried out without the influence of residual hydrogen gas, we did the following [38]: firstly, we checked the contamination rate of hydrogen on a Pt surface in the same experimental setup. On this surface, individual hydrogen atoms can easily be imaged with STM. By using this data, an adsorption rate of 1.3 hydrogen atoms per 1000 nm^2 per hour was found. Since the sticking coefficient S of hydrogen on Fe [47] is lower

	top	side	a (pm)	b (pm)	d (pm)	V (nm^3)
Cu, fcc			255	255	208	0.01353
Fe, fcc			255	255	200	0.01301
			(253)	(253)	(207)	(0.01325)
Fe, bcc			255	282	212	0.01524
			(247)	(285)	(202)	(0.01422)

Figure 4.2: Cu, fcc Fe and bcc Fe lattice parameters as measured for the 2 ML Fe/Cu(111) in comparison with the ideal values given in parentheses. Surface atoms are shown in brighter color, subsurface atoms are darker. The volume per atom can be directly calculated as $V = a \times b \times d$.

($S = 0.1$) than that on Pt ($S = 0.95$), the estimated contamination after one day is only one hydrogen atom per island ($100\,\mathrm{nm}^2$), which excludes any significant effect on our measurements. Secondly, the Fe/Cu(111) sample was intentionally exposed to a dose of $2\,\mathrm{Langmuir}$ hydrogen. After adsorption of hydrogen, the atomic structure of the surface changed: high resolution dI/dz maps show a (2×2) superstructure of the Fe surface (see Fig. 4.3a)) that was not observed on clean samples (see Fig. 4.1b)). Furthermore, dI/dU spectra on the H-covered Fe islands look different from those obtained on clean samples (compare Fig. 4.3b) to Fig. 4.4). Thus, hydrogen-covered surfaces can be easily distinguished from clean surfaces. These findings exclude hydrogen adsorption and desorption as the origin of the later described reversible phase transition.

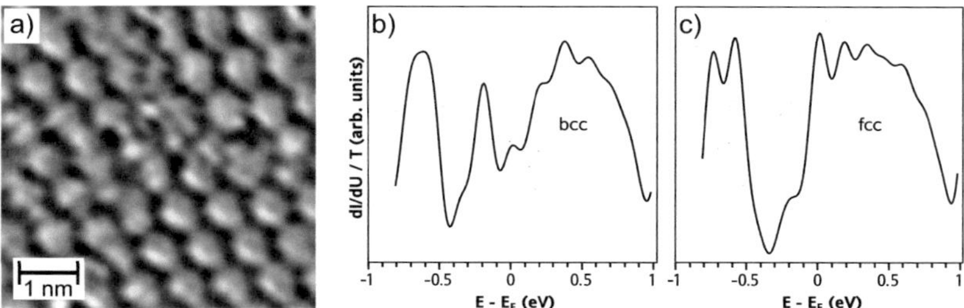

Figure 4.3: a) $3.7 \times 3.7\,\mathrm{nm}^2$ high-resolution STM reveals a 2×2 super-structure on the hydrogen-covered Fe island. b) dI/dU spectra on the two phases of a hydrogen-covered Fe island.

4.3 Magnetic order of 2 ML Fe/Cu(111)

The magnetic order of thin Fe films on Cu(111) has been under debate for quite some time because of the nonlinear increase of the magnetization with increasing thickness and the strong influence of the deposition method on the measured magnetic response [48, 49, 50]. The lack of knowledge of the crystallographic structure lead to multiple hypotheses, e.g. a transition from low-moment ferromagnetic to high-moment ferromagnetic fcc films with increasing thickness. Once the crystallographic structure was revealed by Biedermann et al., they proposed to explain the magnetization measured in thin films by the bcc-like parts of the islands, whereas they attributed no net magnetization to the fcc structure. The available experimental and theoretical data, however, were not sufficient to reveal the magnetic order of the two phases. Therefore, we determined the crystallographic, electronic and magnetic structures of the Fe/Cu(111) system from a comparison of the normalized experimental scanning tunneling spectra (see 2.2.3) and the LDOS as calculated from *ab initio* [38]. The experimental dI/dU spectra were normalized by the transmission T (see 2.2.3). The experimental spectra obtained on the two domains display distinctly different features as shown in Figure 4.4. The LDOS was calculated from first principles using a Green's function multiple-scattering approach [51] within density functional theory in the local spin density approximation (LSDA). The method was specially designed for layered systems by treating adequately semi-infinite boundary conditions [52]. In this theoretical calculation, the LDOS was traced through various structural and magnetic configurations until agreement with the experiment was achieved. The spectrum of the island rim matches well with the theoretical LDOS of an FM bcc structure (see Fig. 4.4c)), and the spectrum of the island center agrees well with the LDOS of a layerwise AFM fcc Fe (see Fig. 4.4b)). The LDOS of other magnetic configurations did not match the experimental observations at all (see e.g. Fig. 4.4a)). In particular, the prominent peak close to the Fermi energy in the fcc spectrum can only be explained by an AFM order. Spin-polarized STM with magnetically coated tips is not applicable to identify the FM and the AFM phase because the two (crystallographically) different phases *per se* exhibit a different electronic structure which cannot be disentangled from a possible magnetic contrast. Our spin-polarized STM experiments showed no lateral magnetic superstructures within the domains, which excludes row-wise antiferromagnetic, triple-q or 120°-Néel structures. In summary, bilayer Fe islands on Cu(111) exhibit a coexistence of AFM fcc and FM bcc phases which can be identified by their characteristic dI/dU spectrum.

4.4 MEC in 2 ML Fe/Cu(111)

The first indications for an influence of the electric field of the STM tip on the sample structure were found in topographic images: the position of the domain boundary changed from one scan line to another and was also different for forward and backward scanning

the same experiment it was also shown that the dI/dU spectrum of the island, which depends on the magnetic structure, changes accordingly. Besides, we have never observed any indications for the existence of additional phases. Therefore, in the following, the two states will be simply distinguished by their apparent height. This allows us to follow the switching with the position of the tip fixed above the area of interest. By recording the apparent height, the state is read out and by varying the applied bias voltage (i.e., the electric field) transitions can be induced. In this way, we can directly follow the transition on a shorter timescale. An example of such an experiment is presented in Fig. 4.5. It can be seen that the phase transition is well reproducible and that the state switches accordingly to the polarity of the applied electric field pulses.

4.4.2 Control of the switching direction

Concerning both possible applications and a thorough understanding of the physics behind the observed switching, it is essential to find out whether the induced phase transition is deterministic. Although the experiment presented in Fig. 4.5 shows that the AFM fcc (FM bcc) state is attained with a positive (negative) electric field, we will see in the following that a wide range of electric fields may lead to switching in both directions. High electric fields are necessary to induce an asymmetry of the transitions from AFM fcc to FM bcc and back. When electric field pulses of $7.5\,\mathrm{GV/m}$ and $50\,\mathrm{ms}$ were used, the following statistics were observed: positive pulses induced the FM bcc to AFM fcc transition in 30 out of 40 trials, but the inverse transition only 2 out of 24 times. Negative electric field pulses clearly favored the AFM fcc to FM bcc transition (22 out of 24) over the FM bcc to AFM fcc switching (0 out of 25) [40]. These statistics show that it is possible to induce an asymmetry by means of the electric field.

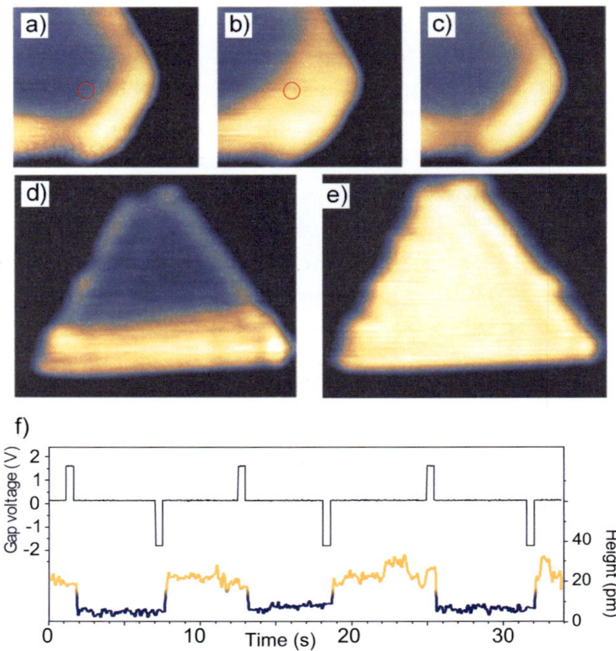

Figure 4.5: a) to c) Three consecutive $6 \times 6\,\text{nm}^2$ STM topographies show switching of a small area of about $1 \times 2\,\text{nm}^2$ on the edge of an island from AFM fcc (a) to FM bcc (b) and back (c). Electric field pulses of $5.5\,^{\text{GV}}/_{\text{m}}$ were applied on the positions indicated with a red circle. A small island can be switched from dominant fcc (d) to full bcc (e). f) Applied voltage ($\propto E$) and apparent height recorded simultaneously.

4.4.3 Revealing the switching mechanism

Although it is shown that it is possible to induce a phase transition with the STM tip, it is not clear *a priori* that the transition is caused by the electric field. When the applied voltage is increased (in order to increase the electric field), also the tunneling current increases, which might lead to changes in the sample structure. In order to understand the origin of the observed switching, a systematic study of the phase transition as a function of the tunneling parameters was performed [38]. The basic experiment was to scan repeatedly along a single line across the domain boundary while the applied bias voltage and the defined tunneling current were varied. First, the applied voltage U was decreased line by line from -0.10 V to -0.25 V with the tunneling current I kept constant. Fig. 4.6a) shows the arrangement of fcc and bcc as a function of the bias voltage plotted on the y-axis. It was found that at a critical value of the bias voltage U_c (equal to -0.16 V in Fig. 4.6a)), a transition from fcc to bcc occurs, shifting the domain boundary to the left. This gives us a value of U_c for the given I. Determination of the critical voltage was then repeated at different tunneling currents, thus different tip-sample distances z, while we were always scanning exactly the same line on the sample. The distance was calculated from the applied voltage and the tunneling current (see paragraph 2.2.4):

$$z = \frac{-\hbar}{\sqrt{2m \cdot \Phi}} \cdot ln(\frac{I}{U} \cdot R_0). \tag{4.1}$$

Finally, a set of points describing the relation between critical voltage U_c and distance z is obtained. These points describe the fcc-bcc domain boundary in the voltage-distance phase diagram. Above the critical voltage, the bcc phase is preferred, and below it, the fcc phase is favored. In the studied interval from 0.3 nm to 0.6 nm, a linear distance dependence of the critical voltage was found; in particular, a freely fitted straight line passes through the origin. Thus, the boundary in the phase diagram corresponds to a constant electric field. A series of experiments with different tips and Fe islands revealed critical electric fields in the range between $3 \cdot 10^8$ GV/m and $9 \cdot 10^8$ GV/m.

To test further for mechanisms not related to MEC, the experimental data were compared with critical voltage-distance relations resulting from different scenarios (Fig. 4.6b)). The transition could be directly caused by the current, for example, due to spin torque [18], spin accumulation [53] or electromigration. This would lead to a switching that only depends on the value of the tunneling current ($I = const.$, green line in Fig. 4.6b)). This, however, disagrees with the observed dependence. Mechanisms that relate to the energy of the tunneling electrons (such as inelastic excitation of specific lattice vibrations or electronic excitations) would result in $U = const.$ (yellow line) but do not fit the data. Similarly, effects that originate from local heating (population of a continuum of excited vibronic or electronic states) do not fit the experimental data ($P = I \times U = const.$, red line). Mechanical forces between the tip and the sample due to the overlap of their wavefunctions [54] that purely depend on the distance between tip and sample, $z = const.$, (blue line) also fail to explain the data. In total, these other switching mechanisms can be safely excluded.

4.4.4 MEC in 2 ML Fe/Cu(111) explained by ab-initio calculations

In order to understand the experimentally observed phase transition induced by MEC, first-principles calculations for the atomic relaxations of the bilayer Fe/Cu(111) surface under an applied electric field were performed with the Vienna Ab Initio Simulation Package (VASP) [55, 56] using density functional theory. An applied electric field of 1 GV/m was used, which is the same order of magnitude as found in the experiments. The STM junction was

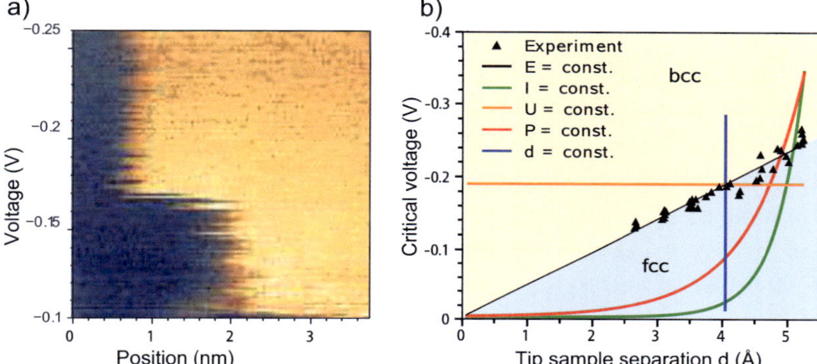

Figure 4.6: Controlling the structure of 2 ML Fe nano-islands with the local electric field. a) A single line across the fcc-bcc domain boundary was scanned with the STM while the voltage U was decreased scan line by scan line. At a critical voltage $(-0.16\,\text{V})$, a transition from fcc to bcc occurs. This experiment has been carried out at different tunneling currents (in a range from $150\,\text{pA}$ to $28\,\text{nA}$), that is, with different tip-sample separations. b) By plotting the critical gap voltage against the corresponding tip-sample separation z, the boundary (black triangles) in the phase diagram is obtained. The experimental results can be compared with possible distance dependencies of the critical voltage (colored lines) resulting from different models (for details see main text): $E = const.$ (black line), $I = const.$ (green line), $U = const.$ (yellow line), $P = const.$ (red line), $z = const.$ (blue line). The curves are all fitted freely to the experimental data.

simulated by a plate capacitor. The results of the simulations of electric-field-induced vertical relaxations are presented in Fig. 4.7a), b) in a cross section of a $[1\,1\,0]$ plane through two Fe layers and two underlying Cu layers. The redistribution of charges is shown as the normalized total charge density shown under the influence of positive (Fig. 4.7a)) or negative (Fig. 4.7b)) electric fields. Here, a positive electric field corresponds to an increase of electron density in the sample. As expected, this leads to a reduction of the topmost interlayer distance (Fig. 4.7a)). The decrease of electron density in a negative electric field (Fig. 4.7b)) results in an increase of the Fe interlayer distance. Based on a total energy analysis, it is predicted that in Fig. 4.7a) the reduced interlayer distances favor layerwise AFM order, whereas in Fig. 4.7b), they are expanded, favoring FM in the system. This is in agreement with the experimentally observed difference of the unit cell volume of FM bcc $(0.0152\,\mathrm{nm}^3)$ and AFM fcc $(0.0130\,\mathrm{nm}^3)$ described in Fig. 4.2 and a general property of Fe: it adopts FM (AFM) order for larger (smaller) unit cells.

As a first step, only vertical relaxations were considered in the calculation. The crystallographic transition from fcc to bcc, however, is mainly determined by a lateral displacement of the Fe surface atoms (see Fig. 4.1). Therefore, also a possible lateral displacement of the Fe atoms was taken into account in the calculations. Fig. 4.7c) shows the two-dimensional energy landscape of the Fe surface atoms as a function of interlayer distance (that can be influenced by the electric field) and lateral displacement from fcc threefold hollow sites to the bcc twofold bridge positions (see paragraph 4.1). Assuming that the energy landscape does not change qualitatively under application of electric fields, we can follow the transition from fcc to bcc as indicated in Fig. 4.7c): an increase of the interlayer distance of the fcc lattice leads to a reduction of the energy barrier towards the bcc lattice. A remaining barrier in the range of some $10\,\mathrm{meV}$ can be overcome by thermal energy (explained in more detail later, see e.g. Fig. 4.8). A relaxation of the lattice follows the potential gradient towards the minimum at the lateral position of a bcc lattice. In a third simulation, both a possible lateral displacement and the electric field were taken into account. Fig. 4.7d), e) show the total energy of the layers for FM (orange) and AFM (blue) configurations. The layers were relaxed along the vertical direction as a function of the lateral displacement. These simulations demonstrate that, under an applied negative electric field (Fig. 4.7d)), the expanded fcc stacking is FM-ordered and unstable and can transform into the energetically favored FM bcc structure. The AFM order is always higher in energy. In the case of a positive field (Fig. 4.7d)), the FM bcc stacking is unstable and fcc stacking with layerwise AFM order of the two Fe layers is energetically most favorable. Furthermore, there is an energy barrier at a displacement of 25 pm for the transformation to the fcc state. In summary, our experimental results are fully reproduced by the *ab-initio* calculations: the applied electric field leads to a vertical deformation of the Fe bilayer which favors either an AFM or an FM order. In combination with the possibility of a lateral deformation that leads to a transformation from an fcc to a bcc lattice, two metastable magnetic states can be attained with the electric field.

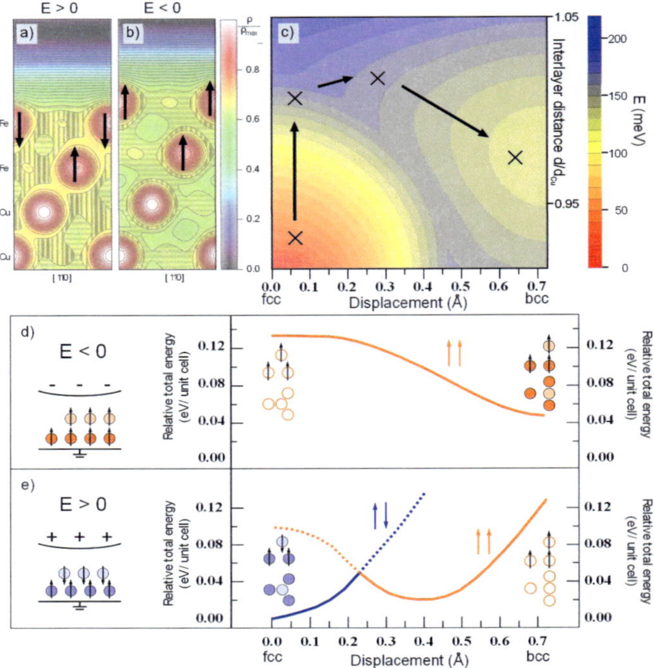

Figure 4.7: Surface relaxations of Fe/Cu(111) in an electric field. a), b) Normalized electronic charge density in 2 ML Fe/Cu(111) under positive (a) and negative (b) electric fields of $1\,\mathrm{GV/m}$. The electric field is modeled by a charged counter electrode placed $0.4\,\mathrm{nm}$ above the surface. The electron charge density is attracted (repelled) by the positively (negatively) charged electrode, causing a vertical compression (expansion) of the Fe bilayer which favors AFM (FM) order. The areas of $0.575\,\rho_{max}$ and $0.6\,\rho_{max}$ are hatched horizontally and vertically, respectively. c) Energy landscape of the top-layer Fe atoms as a function of interlayer distance and lateral displacement shows minimum for the fcc and for the bcc lattice. An electric-field-induced transition from fcc to bcc might follow the path indicated with the black arrows. d), e) Energy landscape for the top-layer Fe atoms as a function of lateral displacement under negative (d) and positive (e) electric fields, simulated for FM (orange) and AFM (blue) order. In a negative (positive) electric field, the lattice adopts the bcc FM (fcc AFM) state and the fcc (bcc) state is unstable. The inset recalls the atomic configuration of fcc and bcc lattices in side and top view. On the left, the corresponding configuration of the STM junction is depicted.

4.4.5 Energy landscape of the phase transition

While it has already been shown that in the case of high electric fields, depending on the polarity, either one or the other phase is found, the dynamic processes that occur at intermediate electric fields will be studied in this part. Both the topographic height, which is related to the tunneling current, and the differential conductance dI/dU can be used to distinguish the two phases. As the dI/dU signal is recorded by using lock-in technique, it shows a better signal-to-noise ratio and is therefore used in the following experiments. The variation in the differential conductance spectra between the two phases (see Fig. 4.4) can be seen when the dI/dU signal is redorded at a fixed z-position of the tip: the interlayer distance is higher in the bcc state and leads to a higher differential conductance. In order to get insight into the metastable regime of the switching, we position the STM tip above the area of interest and apply an electric field such that both states are at similar energy levels and the barrier in between is low enough to allow a reasonable frequency of transitions. Then, the recorded dI/dU signal shows a bistable switching between two clearly different values similar to telegraphic noise (see Fig. 4.8a)). During such a measurement, all tunneling parameters and the tip position were kept constant. After a certain residence time in the AFM fcc state (Δt_{fcc}), it switches to the FM bcc state and stays there for Δt_{bcc} before switching again to the AFM fcc state. In the following, these residence times are the experimental raw data that will be converted to the lifetimes of the two states and to the energy barriers involved in the phase transition. To test for further metastable states, the time-dependent dI/dU signal is plotted in a histogram (see e.g. Fig. 4.8b)). The lateral dimension of the two states can be studied in STM topographies by using very low electric fields that do not affect the lattice (see Fig. 4.8c), d)). It was found that also the lateral size of the domains switched only between two different states. As all measurements discussed in the following have only shown two states, the AFM fcc and the FM bcc state, they will be referred to as fcc and bcc for convenience. The observed random switching at constant applied electric field indicates that the electric field may destabilize the two phases, enabling transitions in both directions while the transition itself is driven by thermal energy. Using Boltzmann's distribution, the ratio of the two populations is directly linked to the energy difference between the two states:

$$\frac{\Sigma(\Delta t_{fcc})}{\Sigma(\Delta t_{bcc})} = \frac{N_{fcc}}{N_{bcc}} = e^{\frac{-(E_{bcc}-E_{fcc})}{k_B \cdot T}}, \tag{4.2}$$

where k_B is the Boltzmann constant, E_{fcc} and E_{bcc} are the energy levels of the fcc state and the bcc state, respectively. In order to find out the lifetime of each state, we plot the N measured residence times as a decay of a population with an initial value of N. Figure 4.8b) shows the decay of the bcc state, with each orange mark representing one measured residence time Δt_{bcc}. The decay is clearly exponential. From a fit, the lifetime of the bcc state (τ_{bcc}) is extracted. In the same way, the lifetime of the fcc state (τ_{fcc}) is obtained. This exponential decay is indicative of an activated transition between two states at local energy minima. The system can be modeled by an energy landscape with the fcc and the bcc state as local minima (see Fig. 4.8c)). In a simple model deduced from transition state theory, it is possible to estimate the absolute values for the barrier seen from each minimum. Therefore, the inverse of the lifetime is converted to the switching rate $\nu = 1/\tau$ which is related to the energy barrier ΔE:

$$\nu = \nu_0 \cdot e^{-\frac{\Delta E}{k_B \cdot T}} \tag{4.3}$$

$$\Leftrightarrow \Delta E = k_B \cdot T \cdot ln(\tau) + k_B \cdot T \cdot ln(\nu_0). \tag{4.4}$$

The offset given by ν_0 is estimated to the order of magnitude of $\nu_0 = \frac{k_B T}{h} = 10^{11}\,\text{Hz}$ [15] at 5 K, corresponding to 10.4 meV. For the example that is shown in Figure 4.8, the lifetime of $\tau_{bcc} = 60\,\mathrm{ms}$ results in a barrier height of $\Delta E_{bcc} = 9.37\,\mathrm{meV}$.

These energy parameters are used to characterize the switching. Their dependence on electric field and lattice strain will be studied in three experiments and finally be summarized qualitatively in Fig. 4.12.

4.4.6 Electric field dependence of the switching at high strain

In a first experiment, the electric field dependence was measured on the fcc domain close to the boundary which is later classified as 'high' strain. We positioned the tip above the sample at a fixed position (feedback loop off) while recording the dI/dU signal and slowly ramping the applied voltage (which in this case is exactly proportional to the electric field). Fig. 4.9a) shows a typical resulting curve, with the voltage ramped from $-0.2\,\mathrm{V}$ to $0.17\,\mathrm{V}$ during 12.1 s. This allows to study the time-dependent switching at different electric fields. The large scale change of the differential conductance reflects the density of states. Besides this change with the applied bias (electric field), two different states of high and low conductance are found, similar to Fig. 4.8a). The lower conductance corresponds to the fcc state, the higher one to the bcc state. Clearly, the switching frequency and the population of each state strongly vary with the applied electric field. In order to obtain better statistics, five different field regimes were distinguished. For each regime, a more precise measurement with constant applied electric field (as in Fig. 4.8a)) was carried out. The measured residence times for each state and the ratio of the two populations are converted to the energy barriers and the energy difference between the two states (see Fig. 4.9b) to f)) as explained above. At 'low' electric fields (see Fig. 4.9d)), both states are found with approximately the same probability, and a switching

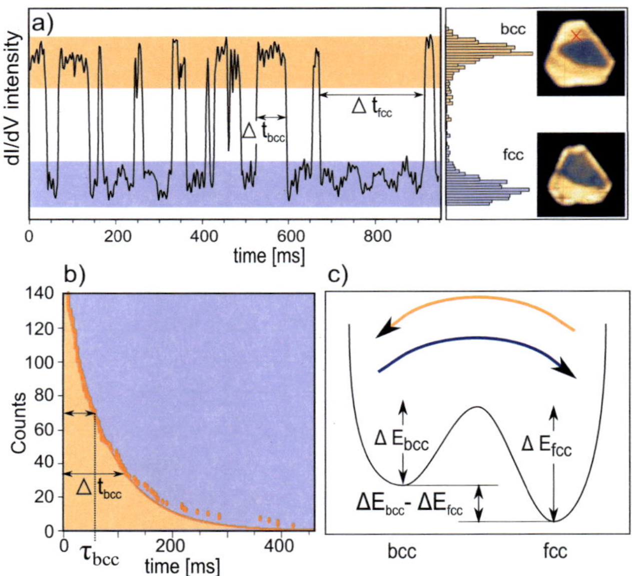

Figure 4.8: Basic experiment to determine the dynamic switching parameters: a) Recorded dI/dU signal at fixed applied electric field and on a fixed position on the island. Examples for residence times Δt_{bcc} and Δt_{fcc} are indicated. The histogram on the right side shows two different intensities and excludes the possibility of intermediate states. b) Exponential decay of the bcc states into fcc states with a lifetime of $\tau_{bcc} = 60\,\mathrm{ms}$. The experimental data (orange dots) can be fitted by a single exponential (orange line). The inset topographies ($11 \times 11\,\mathrm{nm}$) show the lateral extension of the fcc (blue) and bcc (orange) domains within the island. The position where the dI/dU intensity has been recorded is marked by a red cross. c) Energy diagram of the bistable switching between bcc and fcc.

forth and back with low frequency is observed. This corresponds to a small difference between the energy levels of the two states but a high barrier in between them. Increasing the electric field to 'intermediate' values (negative or positive ones) (see Fig. 4.9c) and e)) reduces the barrier between the two states, but keeps them at similar energy. At 'high negative' electric fields (see Fig. 4.9b)), the bcc state is energetically favored, and a high barrier towards the fcc state is found. At 'high positive' electric fields (see Fig. 4.9f)), the fcc state is favored and also separated by a high barrier from the bcc state. This is in full agreement with the proof of principle experiments presented in paragraph 4.4.1 and *ab-initio* calculations (see 4.4.4). Qualitatively, the electric field dependence of the energy landscape can be summarized as follows: with an increase of the absolute value of the electric field, first the energy barrier between the two states is lowered and then, for even higher fields, a difference between the energy levels of the two states opens up. This will be referred to as 'Exp1' in Fig. 4.12.

4.4.7 Strain dependence at high electric fields

The difference in the lattice constants of fcc Fe and fcc Cu(111) substrate (see Fig. 4.2) leads to strain in the pseudomorphically grown islands. As mentioned above, this strain is not uniform, but varies with the position on the island. In the center, the Fe atoms experience the highest strain and adopt the fcc order of the substrate. At the edges, the atoms are able to partially relax and adopt the bcc structure. Close to the domain boundary, the strain is intermediate. Hence, the strain can be varied by measuring at different distances from a domain boundary (only boundaries in $\langle 1\,1\,0 \rangle$ direction were studied (see Fig. 4.1)). Five different regimes are defined in order to qualify the strain: in the central fcc domain the strain is 'very high', close to the boundary at the fcc side it is 'high', on top of the boundary it is 'intermediate', close to the

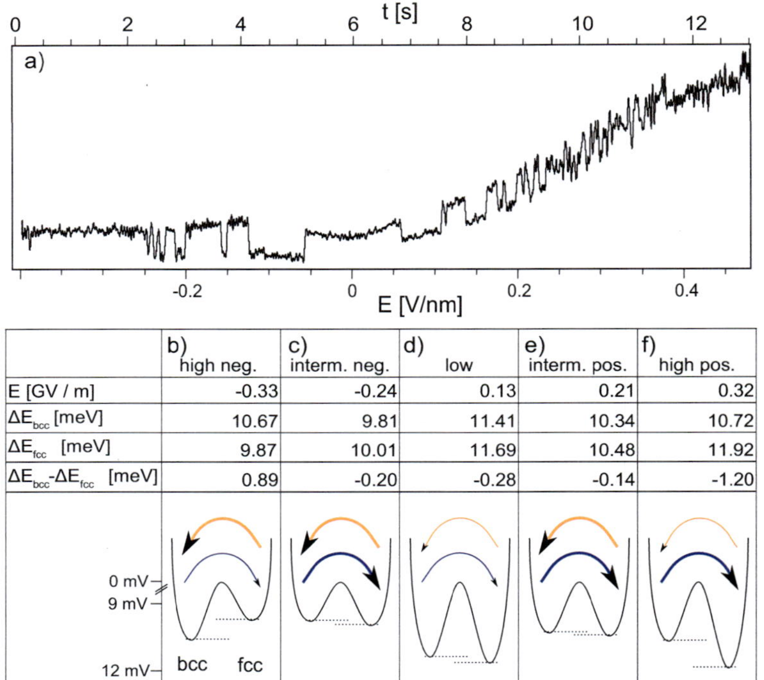

	b) high neg.	c) interm. neg.	d) low	e) interm. pos.	f) high pos.
E [GV / m]	-0.33	-0.24	0.13	0.21	0.32
ΔE_{bcc} [meV]	10.67	9.81	11.41	10.34	10.72
ΔE_{fcc} [meV]	9.87	10.01	11.69	10.48	11.92
$\Delta E_{bcc}-\Delta E_{fcc}$ [meV]	0.89	-0.20	-0.28	-0.14	-1.20

Figure 4.9: Electric field dependence of the switching parameters on the fcc domain close to the domain boundary (later called 'high strain'): a) Typical dI/dU intensity as a function of time and applied electric field. Five different electric field regimes were distinguished, for each of them a more precise measurement of the switching parameters was carried out, shown in b) to f): Energy diagrams for five different electric fields, derived from a total of about 750 transitions. The arrows indicate the different transition probabilities.

boundary at the bcc side it is 'low' and further away it is 'very low'. This method provides a qualitative way of mapping the influence of strain. In a second experiment, we therefore measured the switching parameters with high lateral resolution of the tip position above the island, hence as a function of strain. First, a topographic image of an Fe island at low electric field was acquired. This gives us the native distribution of the two domains and thus the lattice strain (see Fig. 4.10a)). In a second measurement, the electric field was increased to 0.58 GV/m (corresponding to a 'high' field) in order to allow for switching forth and back. On each point of a defined measurement grid, the tip was held at constant distance and at constant voltage while the dI/dU signal was recorded during 18 s. For each point, the residence times of the fcc and the bcc states were extracted. The results can be presented as a map of the relative population of the bcc state, which determines $E_{fcc} - E_{bcc}$, and a map of the switching frequency which is related to the barrier in between (see Fig. 4.10b) and c)). Also $DeltaE_{fcc}$ and ΔE_{bcc} can be mapped (see Fig. 4.10d) and e)). It can be seen that only on a narrow stripe (about 1 nm parallel to the boundary) on the original bcc domain, transitions are observed. Within this bistable area, the variation of the dynamic parameters is relatively small: the energy barriers of the bcc and the fcc states vary between 9.15 mV and 9.84 mV, and between 9.01 mV and 9.71 mV, respectively. The center of the bcc domain and the fcc domain did not show any transition, which corresponds to $\tau \approx 18$ s and barriers of more than 11.74 mV. This shows that the switching strongly depends on the position, i.e., on the strain in the island. From this experiment, the parameters for the energy landscape at 'high' electric field and 'low' strain were deduced. Furthermore, we can conclude that for 'high' and 'very high' strain areas the barrier seen from the fcc side is very high, and for the 'very low' strain area the barrier of the bcc state is very high ('Exp2' in Fig. 4.12).

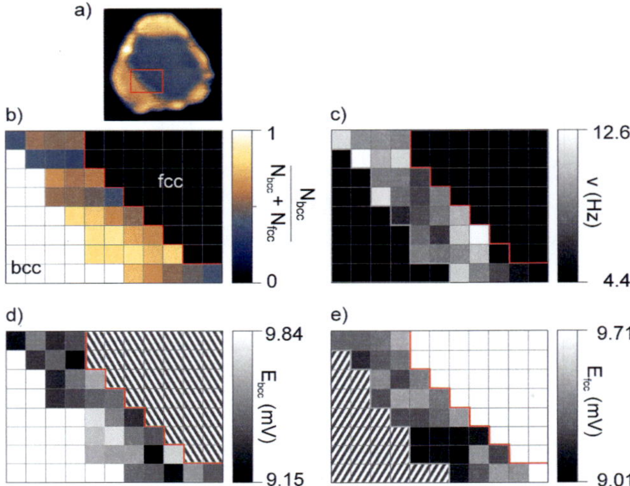

Figure 4.10: Strain dependence of the switching parameters at high positive electric field of $0.58\,\mathrm{GV/m}$: a) STM topography of the Fe island recorded at low electric field gives the original position of the domain boundary. The grid for switching measurements b) to e)) at $0.58\,\mathrm{GV/m}$ (12×8 pixels corresponding to $3.61\,\mathrm{nm} \times 2.24\,\mathrm{nm}$) is marked by a red rectangle. b) Relative population of the bcc state, corresponding to a variation of $\Delta E_{bcc} - \Delta E_{fcc}$ from $-0.17\,\mathrm{meV}$ to $0.72\,\mathrm{meV}$. The fcc domain (black, '0') and the center of the bcc domain (white, '1') were perfectly stable. The original domain boundary is indicated as a red line. c) Switching frequency on the same grid. d), e) Energy barrier seen from the bcc (fcc) state. In the center of the bcc domain (in the fcc domain) this barrier is estimated to be higher than $11.74\,\mathrm{mV}$ because no transition was observed during $18\,\mathrm{s}$. Measurement points that showed no transitions are hatched.

4.4.8 Strain dependence at low and intermediate electric fields

In a third experiment, the electric field dependence of the switching was measured with spatial resolution, i.e., at different strains. A measurement grid of $2.71\,\text{nm} \times 0.78\,\text{nm}$ is centered on the domain boundary (red rectangle depicted in the inset STM topography of Fig. 4.11a)). On each measurement point, the switching was studied as a function of time and applied electric field, similar to the first experiment presented above. A typical curve of the differential conductance (measured on top of the domain boundary) is shown in Fig. 4.11a). For this combination of tip and island geometry, the studied electric field range from $-0.1\,\text{GV/m}$ to $0.1\,\text{GV/m}$ corresponds to 'low' and 'intermediate' fields. It can be seen that, similar to Fig. 4.9a), the switching frequency increases with the increase of the electric field from low to intermediate (pos. or neg.) values. On each point, these differential conductance curves were divided into four equal parts and analyzed as described above. The resulting maps of the population of the bcc state, switching frequency and the energy barriers are shown in Fig. 4.11b) to e). The position of the domain boundary, visible in the bcc population maps, is found not to vary within this range of applied electric fields (Fig. 4.11b)). The average switching frequency, however, increases with increasing absolute value of the electric field (see Fig. 4.11c)). This is in agreement with the observed increase in switching frequency (from 'low' to 'intermediate' electric fields) described in Fig. 4.9. As can be seen in the maps of the switching frequency, the barrier between the two states is lowest (the switching frequency is highest) exactly on the domain boundary which is the central row of the grid. From the maps of the energy barriers (Fig. 4.11d) and e)), the switching parameters for 'low' and 'intermediate' electric fields at 'high', ' intermediate' and 'low' strain were extracted ('Exp3' in Fig. 4.12).

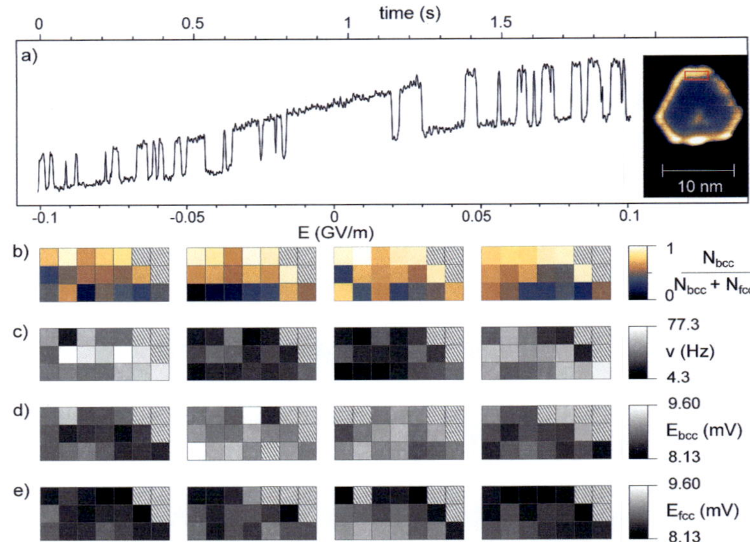

Figure 4.11: Combined electric-field and strain-dependent measurement of the switching parameters. a) Typical dI/dU curve as a function of time and applied electric field. Maps of the relative population of the bcc state (b), the switching frequency (c) and the energy barriers (d) and e)) on a grid of 7×3 pixels (corresponding to $2.71 \times 0.78\,\mathrm{nm}^2$). The maps are placed below the corresponding electric field range in a). The domain boundary is located at the central row of the grid. Measurement points that showed only very few transitions are hatched.

4.4.9 Summary of electric field and strain dependence

The dependence of MEC on the applied electric field and strain in the bilayer Fe islands is summarized in Fig. 4.12. For each combination of strain and electric field, the switching parameters, which are the ratio of the populations of the two states and the switching frequency, are tanslated into an energy landscape as explained at the beginning. The configurations of strain and electric field that (mostly) lead to AFM fcc and FM bcc are shown in blue and orange, respectively. Besides the results from the experiments described above (Exp1, Exp2, Exp3), the behavior at high negative and high positive electric fields at the low strain regime described in paragraph 4.4.2 is included and labeled as R1. The switching of the lattice on top of the domain boundary ('intermediate strain') at 'high negative' electric fields is described in Fig. 4.5 (referred to as R2). Although the crystallographic details of the lattice strain in the two coexisting phases (see Fig. 4.1) were neglected, in summary, the different experiments give a coherent picture of the MEC in 2 ML Fe/Cu(111) as a function of strain and electric field.

4.4.10 Reproducibility of the switching

The reproducibility of the MEC-induced phase transition was confirmed by using the characterization described in 4.8. The dI/dU signal was recorded on a fixed position above the sample with constant applied electric field ($0.8\,\mathrm{GV/m}$) for about $100\,\mathrm{min}$. The time interval is divided into 100 equal parts and for each of them the lifetime was evaluated and converted to the corresponding energy barrier. During the whole measurement, only two different intensities of the dI/dU signal were observed. As can be seen in Fig. 4.13, it took more than 20,000 transitions until the switching characteristics, i.e., the energy barrier changed. This is a surprisingly high number, considering that the crystallographic phase transition

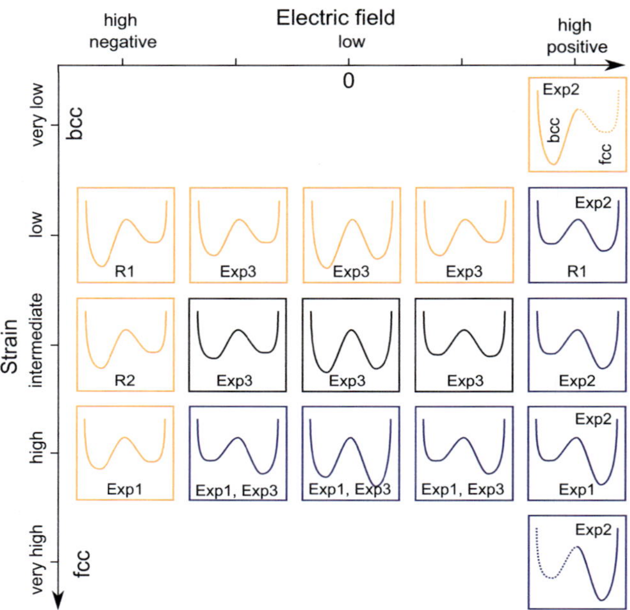

Figure 4.12: Electric field and strain dependence of the energy landscape of 2 ML Fe/Cu(111). The left minimum corresponds to the bcc lattice, the right one to the fcc lattice. Five different electric field regimes and five different strain regimes are distinguished. The energy landscapes qualitatively combine the experiments described above. Configurations that lead to AFM fcc and FM bcc are drawn in blue and in orange, respectively. Dotted parts are guesses following Fig. 4.10b).

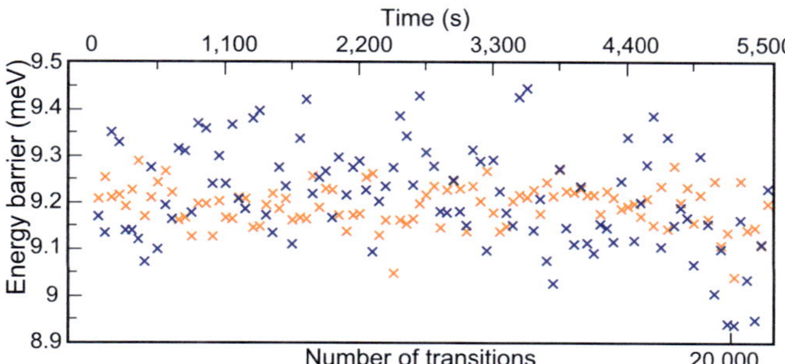

Figure 4.13: Lifetimes of the fcc and the bcc state calculated for repeated switching during $1\,\mathrm{h}$.

involves a displacement of a dislocation.

Switching with short electric field pulses

Fig. 4.14 shows that it is possible to induce a transition even with pulses as short as $60\,\mu\mathrm{s}$ (shorter pulse lengths could not be achieved with our equipment). The black line denotes the bias voltage as a function of time, the colored line displays the z-position of the STM tip measured with a fast oscilloscope. During the pulse of $4\,\mathrm{V}$, the feedback loop was opened and $7\,\mathrm{ms}$ after the pulse it was closed. This is necessary to allow the *I-V* converter to relax (bandwidth $7\,\mathrm{kHz}$). When the feedback loop was closed, the tip moved down $12\,\mathrm{pm}$, which indicated the phase transition. The speed of this movement is limited by the frequency of the feedback loop. It was observed that the transition probability decreases with decreasing pulse width. The absolute value of the electric-field-induced change of the energy barrier between the two states can be estimated from a comparison of the observed stability at low fields (at least in the order of minutes, $\tau \approx 60\,\mathrm{s} \Rightarrow \Delta E \approx 6.5\,\mathrm{meV}$) with the switching during the pulse, i.e., $\tau \approx 60\,\mu\mathrm{s} \Rightarrow \Delta E \approx 12.2\,\mathrm{meV}$.

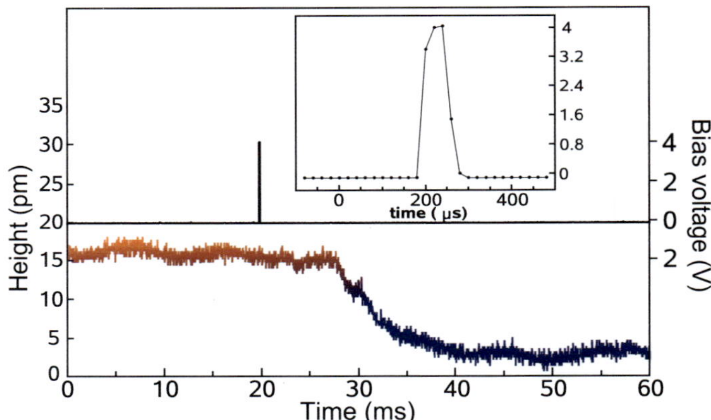

Figure 4.14: Applied bias (black line) and measured height (colored line) at fixed tip position. A pulse of $4\,\mathrm{V}$ and $60\,\mu\mathrm{s}$ was applied (see inset).

5 Experiments on Fe/Ni(111)

Considering the interatomic distances between nearest neighbors d_{nn} and the resulting lattice mismatch as discussed in 4.1, the Ni(111) surface $d_{nn} = 248\,\mathrm{pm}$ should be an ideal candidate to promote the coexistence of Fe fcc ($d_{nn} = 253\,\mathrm{pm}$) and bcc ($d_{nn} = 247\,\mathrm{pm}$) films. Compared to the system Fe/Cu, considerably less work is published on the growth and the crystallographic structure of Fe/Ni. The first layer of Fe/Ni(111) was considered to nucleate in the hcp sites in theoretical calculations and photoelectron diffraction experiments [57, 58], while in other experiments, the fcc sites seemed to be favorable [59]. This fact already indicates a possible coexistence of fcc and hcp phases in the first layers of Fe. With increasing thickness, a transition from fcc to bcc was reported at thicknesses of 3 to 6 ML [60, 58]. A considerable difference compared to the system Fe/Cu(111) is given by the ferromagnetism of the Ni(111) substrate. Altogether, this makes the system of Fe/Ni(111) a complex but promising system in the view of possible MEC.

5.1 Preparation of Fe/Ni(111)

Although similar cleaning techniques as for the Cu(111) surface were applied, obtaining an atomically clean Ni(111) sample was by far more difficult. A Ni single crystal always includes a considerable amount of contaminations in the bulk, e.g. sulfur and carbon. After testing different combinations of sputtering, sputtering at elevated temperatures and annealing to very high temperatures of up to

1200 °C, we finally obtained the best results (about 1 dirt molecule per 100 Ni atoms, see Fig. 5.1a)) with the following preparation procedure: as the annealing not only flattens the surface but also leads to a segregation of contaminations from the bulk to the surface, a high annealing temperature of more than 700 °C was chosen. The last sputtering was rather short (about 30 min), just enough as to remove the residual contamination due to the previous annealing. The final annealing was done at substantially lower temperatures than the previous ones (at about 400 °C) in order to suppress further segregation of contaminants to the surface. For the growth of Fe films on the Ni(111) surface, an optimized preparation procedure that keeps the sample always at the lowest pressure available turned out to be even more crucial. After preparing the Ni(111) surface as described above, the sample had to cool down to room temperature in order to avoid intermixing of the deposited Fe film with the Ni substrate. Although this was done at a pressure of about $1 \cdot 10^{-10}$ mbar (measured by an ion gauge), during this time (about 1 h) the reactive Ni(111) surface became substantially contaminated. These contaminations led to irregular growth of the Fe film (see Fig. 5.1b)) and extended clusters with a 2×2 superstructure on the bare Ni and especially on the surface of the deposited Fe film (see Fig. 5.1d)). One explanation could be hydrogen-induced surface effects as observed for deliberate hydrogen dosing by An et al. [61]. Auger electron spectroscopy revealed a contamination with carbon which could also lead to the formation of a superstructure. The number of adsorbates on the surface could finally be reduced to a minimum by the following measures: 1) The last annealing of the Ni(111) surface was minimized in time to about 1 second at maximum temperature, which led to a reduced heating of the sample surroundings and thus to a shorter cool-down time. 2) Directly after annealing the Ni(111) crystal, it was placed in the STM chamber which has a lower base pressure of residual gases due to the cryostat acting as a cryo-pump. 3) Just before the deposition

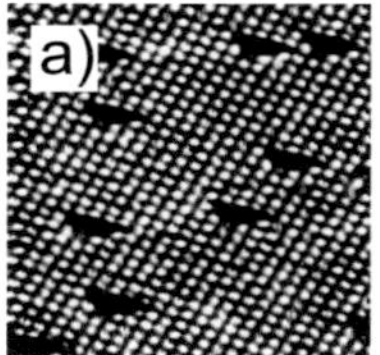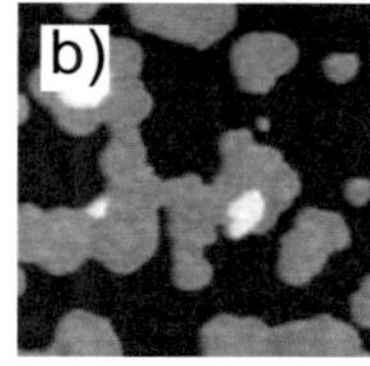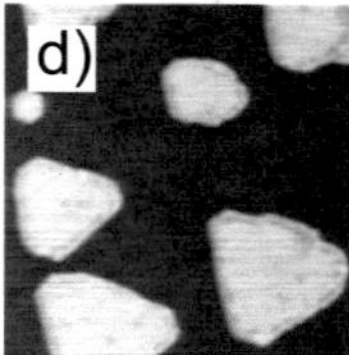

Figure 5.1: a) Atomically resolved STM topography of the bare Ni(111) single crystal surface, $7 \times 7\,\mathrm{nm}^2$. b), d) $60 \times 60\,\mathrm{nm}^2$ topography of similar amounts of Fe deposited at about $30\,^{\circ}\mathrm{C}$. b) Fe/Ni(111) prepared in the normal way, which led to a strongly contaminated surface and irregular island growth. c) $\sqrt{2} \times \sqrt{2}$ superstructure induced by adsorbates on an Fe island (enclosed in red) and on the Ni surface. d) Optimized preparation with reduced surface contamination, for details see text.

of Fe, the pressure in the preparation chamber was improved by sublimation of titanium on the walls cooled with liquid nitrogen (see chapter 3), which led to a pressure of about $4 \cdot 10^{-11}\,\mathrm{mbar}$ during the deposition of Fe. The so-prepared submonolayer amount of Fe/Ni(111) showed a well-ordered morphology with larger Fe terraces and the step edges of the islands oriented along the (110) lattice directions of the surface (see Fig. 5.1c)). Deposition of higher amounts results in Fe films locally 2 and more ML thick, showing the same characteristics as reported in [62] (see Fig. 5.3a)). Thicknesses of the Fe film given in ML correspond to the local thickness.

5.2 Crystallographic structure of Fe/Ni(111)

Regarding previous experiments on Fe/Ni(111) [62], the transition of the Fe lattice from fcc to bcc with increasing thickness seemed to be promising as it could be exploited for MEC via a crystallographic transition similar as in the case of Fe/Cu(111). Therefore,

Fe/Ni(111) films with local thicknesses of 1 to 4 ML were grown. Previous STM studies of the first ML Fe/Ni(111) have shown triangular reconstruction lines along the $\langle 1\,1\,0\rangle$ directions [62]. In the present work, they have never been observed. Instead, in topographic measurements, bright domain boundaries can be seen which enclose irregular domains (see Fig. 5.2a)). Measurements of the differential conductance dI/dV at energies close to the Fe surface state at $-200\,\text{mV}$ [63] reveal the white lines in the topography as boundaries between two electronically different domains: in a dI/dV map taken at $-130\,\text{mV}$ (see inset in Fig. 5.2b)), the Fe monolayer shows a reduced intensity inside the domain. In the dI/dV spectrum in the range from $-1\,\text{eV}$ to $+1\,\text{eV}$, the two phases of 1 ML Fe show different intensities in the negative energy range (see Fig. 5.2b)). In order to identify the crystallographic structure of the two phases, we performed atomically resolved STM measurements. Fig. 5.2c) shows an enlarged view of a 'darker' domain on the right part with atomic resolution. The superimposed hexagonal grid is adjusted to the Fe atoms outside the domain (on the right side of the image) which have a perfectly hexagonal structure and follow the Ni substrate stacking in an fcc order. Within the domain, the fcc grid (blue) does not lie on top of the Fe atoms but exactly in between on threefold hollow sites. This indicates an hcp stacking of the Fe monolayer with respect to the Ni substrate inside the domain. Hence the observed boundaries are surface dislocations between fcc and hcp domains. About 10 % of the first Fe layer shows hcp stacking with domain boundaries mainly perpendicular to the Ni step edges. This is in agreement with photoelectron diffraction measurements [64]. As can be seen in Fig. 5.2c), the introduction of the hcp domains leads to a reduction of atomic densities in the top layer: counting the number of atoms along the $\langle 1\,1\,0\rangle$ direction across the hcp domain gives 1 atom less than expected for the underlying fcc lattice. The observed density of domain boundaries of approximately $(18\,\text{nm})^{-1}$ results in a reduction of the atomic

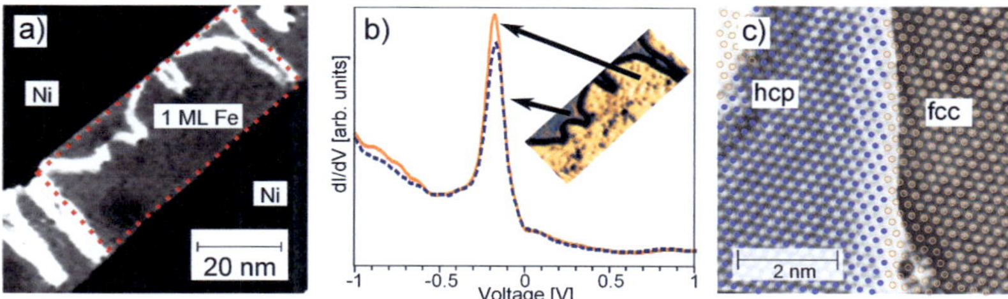

Figure 5.2: Fe/Ni(111): a) STM topography of a 1 ML Fe film (light gray) at a Ni step edge (black). White lines can be seen on the Fe terrace. b) . In a dI/dV map at $-130\,\text{mV}$ (see inset, position indicated by red rectangle in a)), two different intensities divide the 1 ML Fe terrace into domains. dI/dV spectra outside the domain on the Fe terrace (solid orange) and inside the domain (dashed blue) show different LDOS for the two phases. c) Atomic resolution dI/dz map showing a coexistence of hcp and fcc lattices with light domain walls. The overlaid hexagonal grid is adjusted to the fcc lattice and colored orange on the fcc area and blue on the hcp area.

density along the terrace by about $1.4\,\%$. This might be explained by reduction of the strain energy of the fcc Fe lattice as its lattice constant is about $2\,\%$ higher than that of Ni.

As was first published by An et al. [62], starting from the second ML the Fe surface shows a superstructure with periodic stripes aligned perpendicularly to the three closed-packed $\langle 1\,1\,0\rangle$ directions. They observed bright and dark stripes in STM and two satellite spots in the corresponding LEED pattern. Therefore, they interpreted the 2 ML Fe as a bcc $(1\,1\,0)$ lattice in the NW orientation that is expanded in the $[1\,\overline{1}\,0]_{bcc}$ direction by $6.5\,\%$. In our STM measurements, we observed a similar stripe pattern (see Fig. 5.3a)). In dI/dU maps at energies close to the Fe surface state, we could identify three different intensities of the stripes and thus find a doubled periodicity (see Fig. 5.3b)), which is in contradiction to the model proposed by An et al. An atomically resolved dI/dz map of both the Ni surface and the 2 ML Fe film (see Fig. 5.3c)) allows us to precisely determine the atomic lattice: the 2 ML Fe film is expanded by 7.5 % in the $[1\,\overline{1}\,0]_{fcc}$ direction and by 1.9 % in the $[1\,1\,\overline{2}]_{fcc}$ direction with respect to the Ni lattice. Fig. 5.3d) proposes an atomic model that explains the experimentally observed features by a repetition of stripes with fcc, bcc, hcp, bcc-like stripes with a periodicity of 3.2 nm. The fcc and hcp areas appear dark, the bcc-like areas appear bright in the dI/dV-map (compare to Fig. 5.3b)).

5.3 Magnetic structure of 1 ML Fe/Ni(111)

Ab-initio calculations predicted a non-collinear antiferromagnetic order in 1 ML Fe/Ni(111) which could not be observed experimentally. The different crystallographic structure of the fcc and the hcp lattice is expected to lead to slightly different magnetic moments of the Fe film.

5.4 Electric-field-induced hcp-fcc transition in 1 ML Fe/Ni(111)

5.4.1 Bistable switching

In order to test for MEC, high electric fields were applied by increasing the bias voltage. No changes were observed in the stripe pattern of the Fe films thicker than 1 ML. But in the 1 ML Fe film, it is possible to induce structural changes with high electric fields. In Fig. 5.4, the same field of view is shown before (a) and after (b) an electric-field-induced change. The overlaid hexagonal grid is adjusted to the fcc lattice and colored in orange on the fcc area and in blue on the hcp area. It can be seen that a small area of about 20 atoms switched from the hcp domain to the fcc

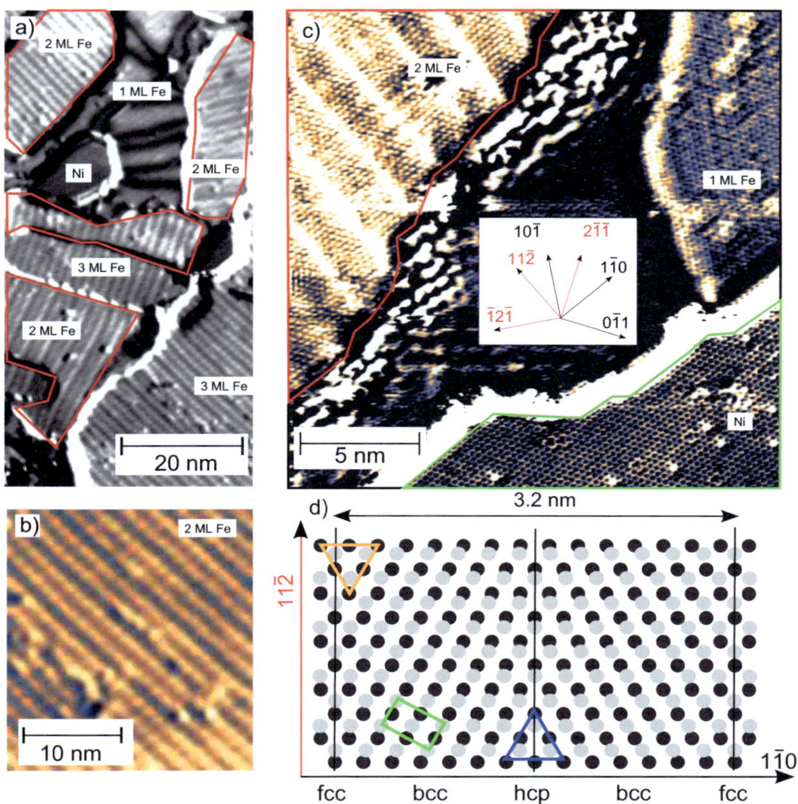

Figure 5.3: a) Fe/Ni(111) films thicker than 1 ML show a stripe pattern perpendicular to the closed-packed $\langle 1\,1\,0\rangle$ directions. The lattice directions in c) are given for the fcc lattice and hold for all STM images presented. The 2 ML Fe films are enclosed with a red line. b) A dI/dV map at $-240\,\mathrm{mV}$ shows three different intensities of the stripes. c) dI/dz map on the bare Ni surface (green), $1\,\mathrm{ML}$ Fe and $2\,\mathrm{ML}$ Fe (red) with atomic resolution. d) Atomic model of 2 ML Fe/Ni(111) for the observed stripe pattern. The relative atomic distances of the top layer (gray circles) are taken from the measurement in c). The expansion of the top layer by $7.5\,\%$ in the $[1\,\overline{1}\,0]$ direction leads to a periodicity of 13 atoms which corresponds to $3.2\,\mathrm{nm}$. The bcc lattice (green rectangle) is characterized by a twofold bridge-stacking (see Fig. 4.1), while the two possible threefold hollow sites correspond to fcc (orange triangle) and hcp domains (blue triangle).

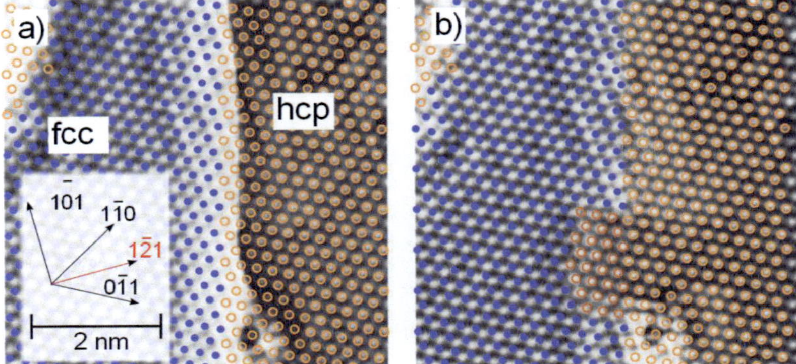

Figure 5.4: Atomic-resolution dI/dz map of 1 ML Fe as shown in Fig. 5.2c) before (a) and after (b) an electric-field-induced change of the lattice. About 20 atoms change their stacking from hcp (open circles, orange) to fcc (closed circles, blue), thus shifting the domain boundary to the left by about 1 nm.

domain, thus shifting the domain boundary by about 1 nm. This transition corresponds to a collective displacement of the Fe atoms by 143 pm along the $[1\overline{2}1]_{fcc}$ direction from the hcp threefold hollow sites to the fcc positions. The new domain boundary in this direction (which generally extends over 3 to 5 interatomic distances) is confined to only one interatomic distance. This corresponds to an increase of the nearest-neighbor distance by 25.8 %.

In a further experiment, the bistability of the switching was confirmed. An hcp domain was imaged at a relatively low electric field of 0.6 GV/m (see Fig. 5.5a)). The domain boundaries appear dark, and the hcp and the fcc domains are marked in blue and in orange, respectively. A high electric field pulse of 7.9 GV/m and 0.1 s was applied at the position marked with the red cross in order to locally change the structure. A second STM image at 0.6 GV/m shows the manipulated configuration with reduced hcp domain (see

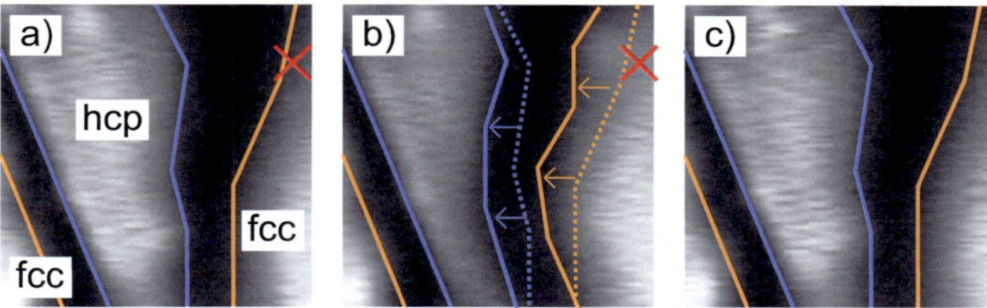

Figure 5.5: Reproducible switching of the crystallographic structure of 1 ML Fe/Ni(111) imaged on $6.25 \times 6.25\,\mathrm{nm}^2$ (dI/dV map at $-190\,\mathrm{mV}$). a) The initial hcp domain is indicated by a blue line, the fcc domain by a orange line, domain boundaries appear dark. An electric field pulse is applied on the position marked with the red cross after the acquisition of the image. The new configuration with a smaller hcp domain is imaged in b) with the initial position of the domains indicated by dashed lines. A second pulse brings the system back to the initial state (c)).

Fig. 5.5b)). After a second electric field pulse, the original situation was restored (see Fig. 5.5c)).

5.4.2 Electric field as the origin of the switching

In order to prove that the observed switching is induced by the applied electric field, we performed a systematic study varying the applied voltage and the distance between tip and sample (see chapter 4.6). For each distance, which can be calculated from the tunneling current and bias voltage, the same line across the hcp domain was scanned repeatedly while the bias voltage was increased. This allows to measure the critical value of the voltage at which the domain boundary is displaced (see Fig. 5.6a)). By repeating this experiment for a set of distances, we obtained a phase diagram of the 1 ML Fe film as a function of applied voltage and distance between tip and sample. The measured critical values correspond to the phase boundary and can be nicely fitted by a linear distance dependence of the voltage as it is expected for a critical electric field. This excludes other mechanisms than the electric field because they would result in distinctly different critical-voltage-distance relations (see chapter 4.6). The negative offset of the voltage for zero distance can be explained by the difference of the work functions of the tungsten tip and the Fe sample surface which induces a potential difference even without an externally applied bias. The slope of the linear fit gives a value of the electric field needed to change the phase of about $10\,\mathrm{GV/m}$ (see Fig. 5.6b)). This value is about one order of magnitude higher than the electric fields needed for MEC in 2 ML Fe/Cu(111). These high electric fields lead to experimental difficulties because a very stable configuration of the tip apex is required for the measurement of the critical voltage. (At the lowest distance, a tunneling current of $2000\,\mathrm{nA}$ was set while the voltage was ramped from $1.5\,\mathrm{V}$ to $3.5\,\mathrm{V}$.)

5.4.3 Dynamics of the switching

Similar to the experiments on Fe/Cu(111), electric fields slightly lower than the critical values lead to switching after a certain residence time. This can be seen in Fig. 5.7a) which shows switching left and right of the domain boundary during scanning at a constant electric field of $7.6\,\mathrm{GV/m}$. However, this implies a high bias voltage (in this case $3\,\mathrm{V}$) which leads to a reduced contrast and a rather low signal-to-noise ratio. Therefore, the distribution of the residence times Δt is studied as follows (see Fig. 5.7b)): the displacement to the left is induced by electric field pulses of $1.4\,\mathrm{GV/m}$ and $0.1\,\mathrm{s}$ (at the positions indicated in the Fig.), while the switching back to the right happens after scanning at $1.0\,\mathrm{GV/m}$ for some time. A displacement to the right has never been observed during scanning at this value. The corresponding histogram of the measured dI/dU intensity at the initial position of the domain boundary shows only

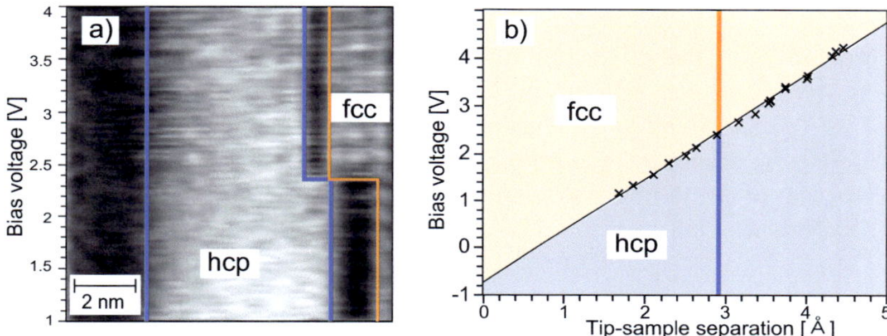

Figure 5.6: Finding the critical parameters for the transition: a) The same line across an hcp domain (blue lines) is repeatedly scanned while the voltage is ramped. Above a certain critical voltage (here 2.3 V), the structure partially changes from hcp to fcc, and the domain boundary (dark) is shifted to the right. Repetition of this measurement at different tip-sample separations results in the phase diagram shown in b). The measurement described in a) corresponds to the dashed blue and the solid orange line giving one data point (black cross). The linear relation between the critical voltage and the tip-sample separation corresponds to a critical electric field of about $10\,\mathrm{GV/m}$.

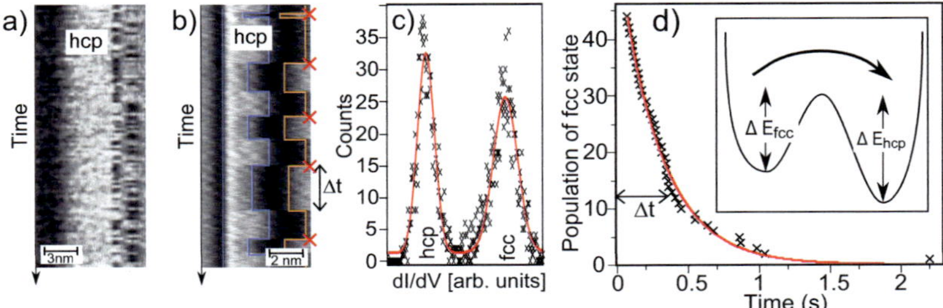

Figure 5.7: a) Switching the domain boundary left and right when scanning across an hcp domain at a constant field of $7.6\,\mathrm{GV/m}$ (dI/dU at $3.7\,V$). b) Switching left by field pulses of $1.4\,\mathrm{GV/m}$ and right by scanning at $1.0\,\mathrm{GV/m}$ across an hcp domain (blue) (dI/dU at $-240\,\mathrm{mV}$). c) In a histogram of the differential conductance, only two stable states are found. d) Exponential decay of the hcp state with a lifetime of $0.3\,\mathrm{s}$.

two different peaks, corresponding to two stable positions of the domain boundary. The plot of the residence times of the hcp state is depicted in Fig. 5.7d) and shows an exponential decay with a lifetime τ_{hcp} of about $0.3\,\mathrm{s}$. Similar to the experiments on Fe/Cu(111) 4.4.6, this can be explained by a thermally activated transition between two local energy minima (see inset in Fig. 5.7d). A clear relation between the polarity of the electric field and the resulting lattice state could not be found. Switching in both directions was possible with pulses of any polarity with approximately the same probability.

6 Experiments on Ni/Cu(100)

Generally, strong MEC is expected for systems which are close to an instability of their magnetic order. An ideal candidate would be a magnetic thin film at the spin reorientation transition (SRT) (see chapter 2.1.1). It is well known that in the system of Ni/Cu(100), two SRTs take place with increasing thickness of the Ni layer. The first one is a transition from in-plane to out-of-plane easy magnetization direction in the range from about 7 ML [65, 66] to 10 ML [67], and the second one rotates the easy magnetization direction back to in-plane at a thickness of about 50 to 60 ML [68].

In the following, we will concentrate on the first transition from in-plane to out-of-plane easy magnetization direction. It arises from the high negative surface anisotropy of Ni. This is why, for very thin films, the negative surface anisotropy is dominant, leading to an in-plane easy magnetization direction. It competes with the magnetoelastic anisotropy that favors out-of-plane magnetization. This magnetoelastic anisotropy is induced by the tetragonal distortion of the Ni film due to the lattice mismatch of about 2.5 % between Ni and Cu. The sum of the different magnetic anisotropies of the Ni/Cu(100) film is given by (see paragraph 2.1.1)

$$u = [\frac{1}{2}\mu_0 M^2 - B_1\epsilon_{11}(1+2c_{12}/c_{11}) - \frac{K_s + K_{int}}{t}]cos^2\theta + K_1 cos^2\theta sin^2\theta.$$

$$(6.1)$$

As the surface anisotropy energy is thickness-independent and the magnetoelastic anisotropy energy is proportional to the volume (i.e., to the thickness; Ni grows pseudomorphically up to 13 ML), above a certain thickness the magnetoelastic out-of-plane component dominates. The critical thickness for this transition is:

$$t_c = 2(K_s + K_{int})/[\mu_0 M^2 - B_1 \epsilon_{11}(1 + 2c_{12}/c_{11})]. \qquad (6.2)$$

The temperature dependences of the anisotropy constants and of the magnetization itself lead to the complex phase diagram shown in Fig. 6.1 (from [69]).

A detailed study of this SRT is presented in [70] and citations therein. The local three-dimensional magnetization vector was measured during film growth with spin-polarized low-energy electron microscopy (SPLEEM) by Farle et al. [71]. The SRT from in-plane to out-of-plane magnetization was found to take place via a phase transition of first or second order depending on the substrate topography: a pronounced step-bunching leads to an abrupt change of the magnetic structure, while a homogeneous step density promotes a continuous SRT [72]. In another experiment, it was shown that the SRT can be triggered by controlled adsorption and desorption of H_2 on the Ni surface [73]. It was found that the surface anisotropy is changed, while the average film strain which determines the magnetoelastic anisotropy remains constant [73]. These experiments suggest a potential influence of the electric field on the surface anisotropy of the Ni film and thus the possibility of an electric-field-induced SRT.

In this thesis, the SRT in Ni/Cu(100) was studied with three different techniques: Firstly, electric fields were applied through a thick organic insulator, and the magnetism of the Ni film was measured with MOKE. Secondly, the Ni film was used as one of two electrodes in a magnetic tunnel junction (MTJ), and its magnetization was derived from changes in the magnetoresistance of the junction. These two studies were carried out in Osaka, Japan, in the group of Prof. Y. Suzuki. This group has already demonstrated the influence of an electric field on the anisotropy of ultra-thin films of Fe, FePd and FeCo [74, 75, 76] measurements. Thirdly, the system was studied with STM.

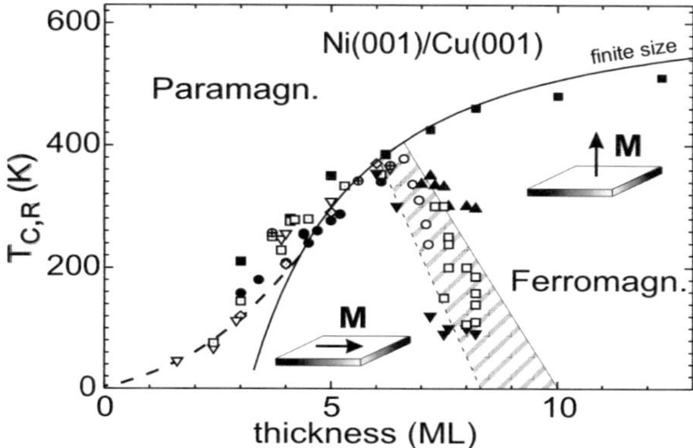

Figure 6.1: Phase diagram of Ni/Cu(100) from [69]

6.1 Polar MOKE measurements

Previous experiments on MEC in Fe and Fe alloys [74, 75] have shown that MOKE (see 2.1) is a suitable tool to study the influence of an electric field on surface magnetism.

6.1.1 Sample preparation

The samples used for the MOKE measurements are composed of a pile of several metallic layers deposited on a $20 \times 10\,\mathrm{mm}^2$ MgO wafer as depicted in Fig. 6.2a), b). All depositions were performed at room temperature. The substrate was first cleaned by degassing at about $500°C$ in UHV for several hours, and a few nanometers of MgO were deposited by MBE in order to obtain a clean surface. Then a 2 nm thick layer of Cr was deposited to increase the adhesion of the other metallic layers. It was followed by deposition of 20 nm of Pd which served as a buffer layer that reduces the lattice mismatch between the substrate and the succeeding 20 nm thick Cu film. This Cu film served as the substrate for the Ni film which was prepared in a wedge shape with a linear thickness variation from 2.1 to 16.3 ML (second sample: 7 to 14.1 ML) and was capped with 2 nm MgO (see Fig. 6.2a)). Coverage with MgO was not necessary for the MOKE measurements but allowed a better comparability with the samples for the TMR measurements. The actual insulating barrier through which the electric field was applied was a 1500 nm thick spin-coated layer of polyimide. On top of the polyimide film, an array of circular junctions of indium-tin oxide (ITO) (1 mm in diameter) was deposited. These junctions were contacted by further deposition of $1 \times 1\,\text{mm}^2$ Au patches which were slightly off-centered in order to keep the ITO junctions partly uncovered (see Fig. 6.2b)). The Au patches were then contacted by a standard bonding technique. In the following, thicknesses of the Ni film are given in ML and correspond to the mass equivalent of a perfectly flat film.

6.1.2 Crystallographic and magnetic structure

The Cu film of 20 nm was expected to be sufficiently thick to attain the lattice constant of bulk Cu. Ni/Cu(100) is reported to grow pseudomorphically up to a thickness of 13 ML [77]. In the RHEED measurements of the Ni growth, a periodic oscillation of the signal indicated well-ordered layer-by-layer growth (see Fig. 6.2c)). The growth of MgO/Ni(100) goes along with a large lattice mismatch of about 16 % (which might be reduced by a 45° rotation of the lattice as in the case of MgO/Fe). Therefore, the quality of the interface and the MgO barrier itself was questionable [78]. Our measurements, however, showed a clear TMR effect through this barrier, which indicated at least a reasonable quality of the interface (see section 6.2.2).

The magnetic structure of the Ni wedge was investigated by polar MOKE measurements that probe the out-of-plane magnetization (see paragraph 2.1). Fig. 6.3a) shows the variation of the intensity

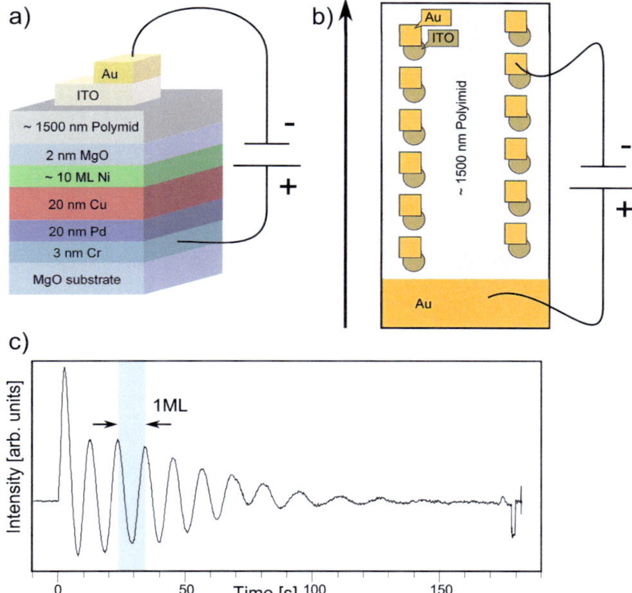

Figure 6.2: a) Stacking of the multilayer grown by MBE and the spin-coated polyimide as it was used for MOKE measurements. b) Top view on the $20 \times 10\,\mathrm{mm}$ sample with the thickness of the Ni layer increasing along the arrow. Each Au/ITO junction of $1 \times 1\,\mathrm{mm}$ covers a different Ni thickness. c) RHEED oscillations of the growth of Ni/Cu.

of the MOKE signal across the sample under an applied magnetic field of 3000 Oe. The thickness of the Ni wedge is indicated by the black line. The MOKE signal decreases almost linearly with the thickness and vanishes at a thickness of 7.8 ML. The latter can be explained by the decrease of the Curie temperature below 300 K with decreasing film thickness. For a 8 ML thick Ni/Cu(100) film, the Curie temperature is found to be 360 K [69].

In order to find the critical thickness for the SRT, we measured hysteresis curves at different positions across the wedge-shaped sample, i.e., at different thicknesses of the Ni film (see Fig. 6.3c)). With decreasing thickness, the saturation magnetization as well as the coercive field decreases. Starting from a thickness of 12 ML, the coercive field drops off rapidly, which indicates the begin of the SRT (see Fig. 6.3b)). Furthermore, the hysteresis curves below 12 ML start to exhibit a finite slope (hard-axis magnetization). From these measurements, the SRT of the MgO-capped Ni/Cu(100) film is estimated to take place in the thickness range between 10 ML and 12 ML. This is a shift of about 1 ML [67] compared to vacuum/Ni/Cu(100) (4 ML when compared to [65]). It is most likely induced by the MgO capping which both changes the surface anisotropy (MgO/Ni interface instead of vacuum/Ni) and the magnetoelastic anisotropy by inducing additional strain. In order to keep a good comparability to the sample for the TMR measurements, the film for the MOKE measurements was also mildly annealed to 200 °C. Annealing reduces the interface roughness and thus is expected to favor out-of-plane magnetization [79]. This was confirmed experimentally for the case of Ni/Cu(100) as an increase of the critical thickness of 0.7 ML upon annealing [80]. In our case, however, the annealing lead to a reduction of the critical thickness of about 1 ML. This might be explained by a partial release of strain at the Ni/MgO interface which leads to a reduction of the strain-induced out-of-plane anisotropy.

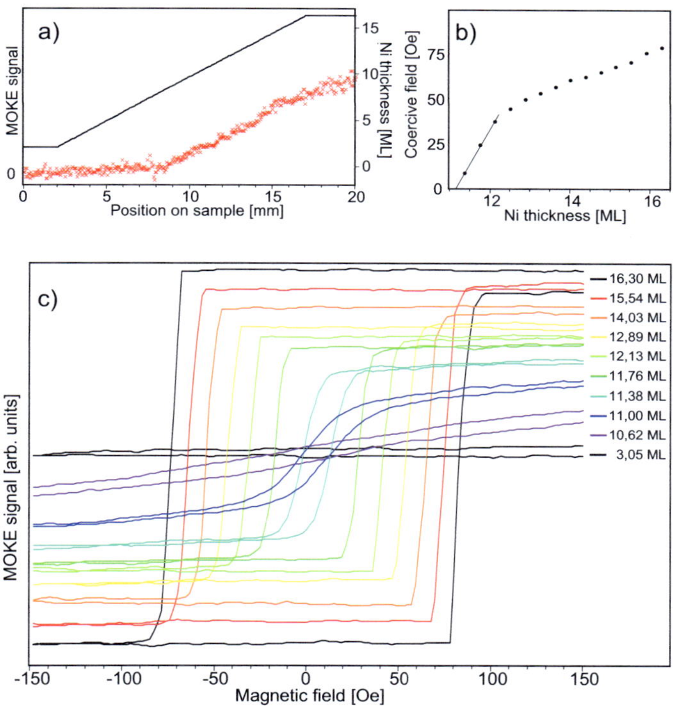

Figure 6.3: a) Cross section of the Ni wedge: thickness calculated from the deposition rate (black line) and saturation magnetization measured with polar MOKE in a magnetic field of $3000\,\mathrm{Oe}$ applied perpendicularly to the surface (red crosses). At $7.8\,\mathrm{ML}$, the MOKE signal vanishes. b) Coercive field as a function of the thickness of the Ni layer. c) Hysteresis curves measured with polar MOKE at different Ni thicknesses.

6.1.3 MEC in 10 ML Ni/Cu(100) studied with MOKE

In order to detect electric-field-induced changes of the magnetic order in the Ni film, hysteresis curves were recorded for different electric fields applied through the polyimide film. The maximum electric field at the Ni surface was estimated by the voltage drop of 210 V through the insulator and the relative dielectric constant of polyimide of 3.4: $E \approx 0.048\,\mathrm{GV/m}$. As the charge distribution at the interfaces MgO/Ni and polyimide/MgO are unknown, this is just a rough approximation and in the following, the applied voltage is given instead of the electric field. Various thicknesses of the Ni film were studied by contacting different Au/ITO junctions along the wedge. Usually, at each thickness several hundred hysteresis loops were recorded. For each loop, the magnetic field was swept from $+200\,\mathrm{Oe}$ to $-200\,\mathrm{Oe}$ and back within 90 s. In Fig. 6.4, a typical measurement for a 13 ML thick Ni film is depicted. The voltage was varied between $-210\,\mathrm{V}$ and $+210\,\mathrm{V}$, and every 30 V a hysteresis loop was recorded. Fig. 6.4a) shows that the variation of the hysteresis loops with voltage is smaller than the line width. In order to characterize the hysteresis, the coercive field was extracted for each loop. Fig. 6.4c) shows the variation of the coercive field as a function of the voltage applied to the junction. Obviously, it slightly varies with the applied electric field. Such a small variation was observed for a wide range of the Ni film thickness and is reported for similar systems measured in an equivalent setup (personal communication). The saturation magnetization remained constant within the experimental errors.

Close to the critical thickness of the SRT, the oscillations of both the coercivity and the saturation magnetization with the applied electric field become significant. Fig. 6.5 shows the variation measured at three slightly different thicknesses (within one Au/ITO junction) close to the critical value. For each thickness, the hysteresis loops are averaged to one period, and the values for the

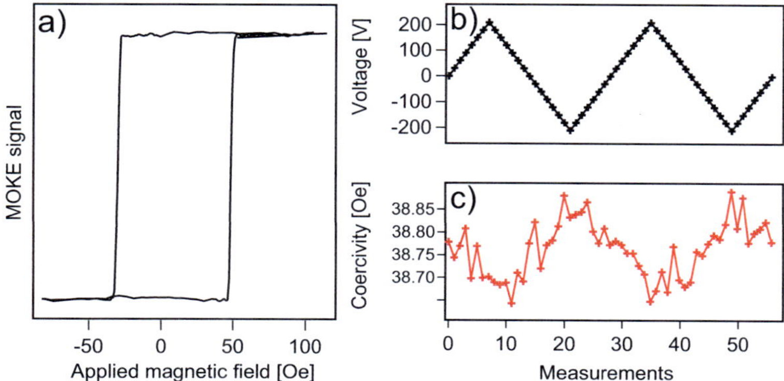

Figure 6.4: Influence of an applied electric field on the coercive field of 13 ML Ni/Cu(100). a) Hysteresis loops are almost identical for all voltages. b), c) The coercivity varies with the applied electric field by about 0.6 %. The coercive field variation is averaged to show two periods of voltage variation.

coercivity (H_c) and the saturation magnetization (M_S) are shown for two periods. Both H_c and M_S clearly vary with the applied electric field. The fact that the maxima of H_c and M_S are slightly out of phase with the voltage variation indicates an effect due to accumulation of trapped charges in the insulator [81]. This could also explain the small hysteresis observed in the coercivity and the remanence.

The observed dependence of MEC on the position on the sample, i.e., on the thickness of the Ni layer, is shown in Fig. 6.6. A variation of 107 % in the coercivity with the electric field (from negative to positive electric field) and a variation of 6.7 % in the saturation magnetization were found for the smallest Ni thickness (10 ML). In comparison, changes of only 10 % in the coercivity and of 1.89 % in the saturation magnetization were observed for the thickest film (10.1 ML). The thickness of 10 ML was estimated

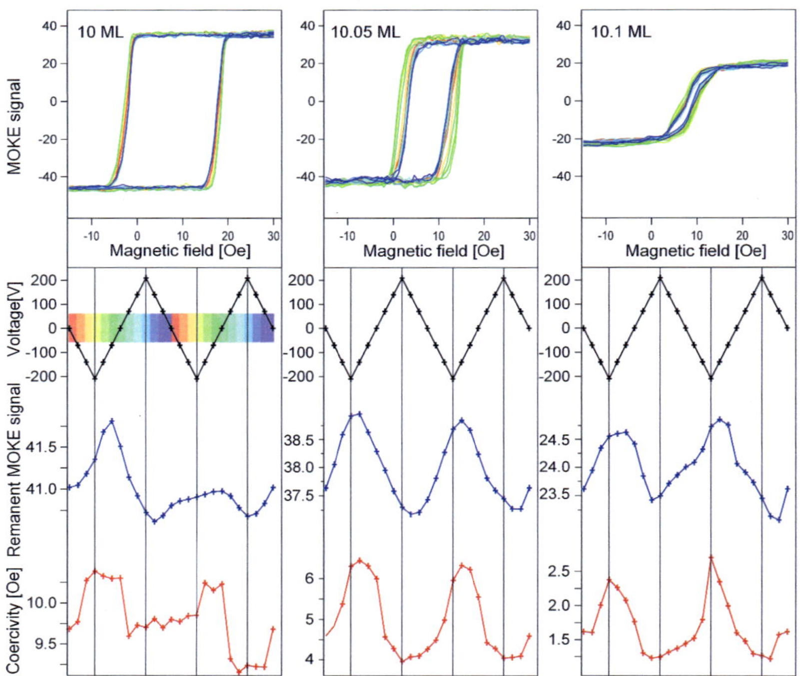

Figure 6.5: Influence of an applied electric field on the hysteresis curves of different thicknesses of the Ni film: a) 10 ML, b) 10.05 ML, c) 10.1 ML (assuming a linear wedge). The hysteresis loops measured with polar MOKE were averaged to one period of voltage variation from 0 to 210 to −210 to 0 V (see spectral color code). Variation of saturation magnetization (blue) and coercivity (red) are shown for two periods.

from the position of the Au/ITO junction on the sample. The relative thickness variation is calculated from the displacement by the stepping motors under the assumption of linear increase of Ni film thickness along the wedge. Although the SRT could not be completely controlled by the electric field, the rather abrupt increase of the effect when the critical thickness is approached (an increase of about 1 % in thickness strongly changes MEC) indicates an effect closely related to the SRT. Although absolute values of the magnetization are unknown, the variation of M_S (H_c) can be related to an increase of the magnetic moment (an increase of the perpendicular anisotropy) under application of a negative electric field (negative voltage applied to Au/ITO electrode). The maximum change of the coercivity of 2.35 Oe is equal to a change of the thickness of 0.07 ML: it can be approximated from Fig. 6.3b) that the increase of coercivity with Ni thickness is about 30 Oe/ML. In order to check for a possible MEC-induced switching of the magnetism, in a further experiment the sample magnetization was first saturated, and then the MOKE signal was recorded at zero magnetic field while the voltage was varied. This experiment was repeated with the sample magnetized in opposite direction. The difference of the two obtained curves is then proportional to twice the saturation magnetization. Fig. 6.6c) shows the difference of the MOKE signals averaged over three periods. It clearly shows an oscillatory behavior indicative of a canting of the magnetization direction due to the applied electric field (see Fig. 6.6). As no hard-axis magnetization hysteresis loops were measured, values of the anisotropies and their changes with applied electric field could not be derived.

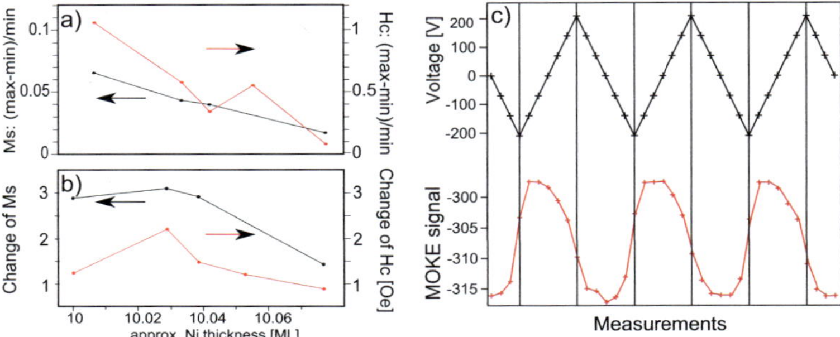

Figure 6.6: Comparison of MEC in Ni films for four different thicknesses close to the SRT: a) relative and b) absolute changes of the coercivity and the saturation magnetization. c) Polar MOKE signal recorded without applied magnetic field as a function of the applied voltage. The thickness of the Ni film is about 10 ML.

6.2 TMR measurements of Fe/MgO/Ni/Cu(100) tunnel junctions

The observation of MEC in magnetic tunnel junctions (MTJs) was first reported for MgO/FeCo junctions in [76]. The recent success of a change of magnetization direction in micro-fabricated MTJs which persists at room temperature signifies an important step towards possible applications of MEC [13].

6.2.1 Sample preparation

The same stacking of metallic layers as for the MOKE samples (see Fig. 6.2) was deposited by MBE. In case of the sample designed for TMR measurements, 20 nm of Fe were deposited on top of the MgO barrier. In order to prevent oxidation, the Fe film was capped by a few nanometers of Au. As the MgO layer always exhibits a certain number of pinholes, it is necessary to reduce the size of the tunnel junctions to about $1\,\mu\mathrm{m}^2$. This ensures at least a certain number of pinhole-free junctions. Therefore, a standard electron beam lithography process was used: after spin-coating the whole sample with a special resist, certain areas were marked by scanning with an electron beam. Only these chemically modified areas could withstand the consequent chemical removal of the resist. The such prepared mask was used to protect the marked areas during the consequent Ar-etching which removed the uncovered metallic layers. With this method, the sample was prepared as described in the following (simplified): in a first step, the bottom electrode was shaped by removal of the metallic layers down to the MgO substrate (green area in Fig. 6.7a) was etched). Next, the actual TMR junction was prepared by etching the Au and Fe layers except for a pillar of $1\,\mu\mathrm{m}^2$ (see arrow in Fig. 6.7b)). Then, a layer of SiO_2 was deposited in order to insulate the Fe top electrode of the pillar from its surroundings. Finally, the top of the pillar was contacted by a Au bridge (Au was deposited on the yellow area in Fig. 6.7c)). In total, an array of 24×12 MTJs was fabricated on the $20 \times 10\,\mathrm{mm}^2$ sample.

6.2.2 Magnetoresistance curves of Fe/MgO/Ni/Cu(100) tunnel junctions

The resistance R of the MTJs prepared as described above was measured at room temperature while an in-plane magnetic field was periodically ramped from $H_{max} = +3000\,\mathrm{Oe}$ to $H_{min} = -3000\,\mathrm{Oe}$. At low voltages of typically $5\,\mathrm{mV}$ (at zero magnetic field), a resistance in the range of $10\,\mathrm{k}\Omega$ was measured for about $20\,\%$ of all junctions. These junctions exhibited a TMR of a up to $8\,\%$ at room temperature and were used for further experiments. (Giant TMR values at room temperature [2, 82] result from a symmetry-dependent filtering of electrons which is found in MgO(001) barriers in combination with bcc-ordered electrodes [83].) The other MTJs

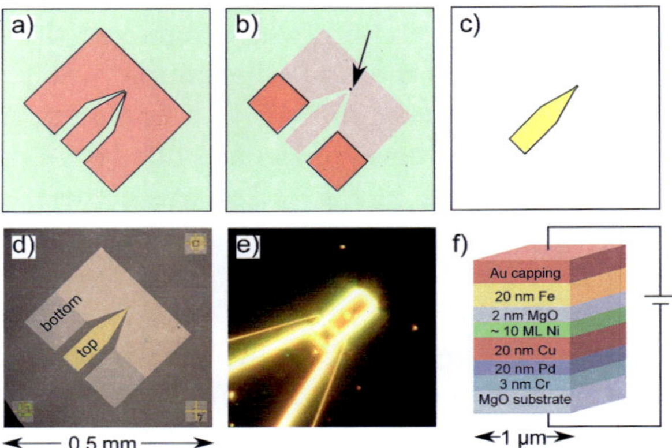

Figure 6.7: a) to c) Three main steps of preparation of an MTJ by lithography. d) Optical microscopy of a single MTJ as prepared by EBL: the gray contacts are connected to the bottom electrode, the contact in the middle (Au, yellow) is bridged to the top of the pillar-shaped TMR junction. e) Dark field microscopy of the pillar (about $1\,\mu\mathrm{m}^2$ in diameter) covered by the Au bridge. f) Stacking of the MTJ with the Ni wedge as bottom electrode and a $20\,\mathrm{nm}$ Fe layer as top electrode. Voltage is applied to the top electrode.

had clearly different resistances or a shorted pillar and did not show a TMR effect at all. Fig. 6.8a) exemplarily shows recorded R-H curves for three different thicknesses of the Ni layer. Generally, thicker Ni films lead to slightly higher resistances. As R-H measurements with the magnetic field applied out-of-plane were not possible, the interpretation of the presented measurements was difficult. As an estimation, the magnetic structure of the Ni wedge was assumed to be identical to the simultaneously prepared MOKE sample, and an SRT at a thickness of about 10 ML was assumed. The easy magnetization direction of the Fe layer was expected to be in-plane. This allows the following interpretation of Fig. 6.8a): the 13 ML thick Ni film exhibits an out-of-plane easy axis of magnetization. Thus, the resistance of the MTJ decreases continuously with increasing magnetic field which rotates the magnetization of the Ni layer into the film plane. The R-H curves for H_{max} to H_{min} and for H_{min} to H_{max} are almost identical, which further corroborates the assumption of an out-of-plane easy axis of magnetization. The maximum applied magnetic field could not saturate the resistance, i.e., not rotate the magnetization fully into the film plane.

According to the MOKE measurements, thinner films (9 ML and 7 ML in Fig. 6.8a)) exhibit an in-plane easy axis of magnetization. The resistance maximum in the absence of an applied magnetic field indicates an antiparallel alignment with respect to the magnetization of the Fe film. This is possibly attributed to magnetostatic coupling of the Ni film to the Fe electrode. After a small plateau at low magnetic fields, the resistance drops once the applied magnetic field overcomes this coupling. For high magnetic fields, the two electrodes are aligned almost in parallel, and the resistance saturates at a low value. Upon closer examination, one finds characteristics of both in-plane and out-of-plane magnetization directions in the magnetoresistance curves: the out-of-plane component of the magnetization leads to a contribution that is symmetric (gray area in Fig. 6.8d)), while the in-plane component gives rise to a fully antisymmetric contribution to the resistance curve (red area for a sweep from H_{max} to H_{min}, $|H| < 500\,\mathrm{Oe}$). As the maximum magnetic field was limited to $+3000\,\mathrm{Oe}$, the TMR ratio of the samples showing strong out-of-plane anisotropy could not be evaluated, and a comparison of the TMR ratio for different thicknesses of the Ni film is omitted.

6.2.3 MEC in 7-13 ML Ni/Cu(100) in MTJs

In order to study the influence of an electric field on the magnetism of the Ni film, we measured the R-H curves at different applied voltages. As a change of the applied bias voltage strongly influences the tunneling resistance and the TMR ratio (see Fig. 6.8b)), the R-H curves were normalized by the maximum and the minimum values. This does not affect a possible influence of the applied bias on the anisotropy: we assumed perpendicular (parallel) orientations of the two magnetic layers at maximum (minimum) resistance for all voltages. Thus, the relative value of the conductance is proportional to the relative orientation of the magnetization directions, which will be explained below.

Fig. 6.8b) shows the positive magnetic field range of magnetoresistance curves (of an MTJ with an 11 ML thick Ni layer) measured at different voltages. It can be seen that the different electric fields lead to different slopes of the magnetoresistance curves. One way to estimate the MEC-induced change in the magnetic order is to calculate the magnetic anisotropy energy. In this particular experimental setup, we have a perpendicular easy axis of magnetization and a magnetic field applied in-plane, so we can only evaluate the in-plane component of the magnetization as a function of the applied magnetic field (called hard-axis magnetization). This allows us to estimate the out-of-plane anisotropy energy. To do so, it is assumed that firstly, the spin polarization is proportional to the magnetization, secondly, the maximum of the resistance (at zero magnetic field) corresponds to a perpendicular alignment of the magnetization directions, and thirdly, the minimum at high magnetic fields corresponds to parallel magnetization directions. Then, the conductance following Slonzewski's rule (see section 2.1.3) is given by

$$G = G_{perp} + \Delta G \cdot cos\theta = G_{perp} + \Delta G \cdot \frac{M_{in-plane}}{M_s}, \qquad (6.3)$$

where G_{perp} is the conductance for perpendicular magnetization directions and $\Delta G = G_{parallel} - G_{perp}$ is proportional to the product of the spin polarizations of the two electrodes. As the absolute value of the saturation magnetization M_s is unknown, only relative changes of magnetization and anisotropy can be extracted.

After normalization of the conductance curves ($G = 1/R$) to a maximum of $G_{parallel} = 1$ and a minimum of $G_{perp} = 0$, the in-plane magnetization is directly proportional to the conductance: $G = \frac{M_{in-plane}}{M_s}$. The anisotropy can then be calculated as the integrated hard-axis magnetization as a function of the applied external field: $E_{perp} = \int M_{in-plane} dH$ (gray area in Fig. 6.8d)). The anisotropy increases approximately linearly with the (positive) applied electric field as it is shown in Fig. 6.8c).

A variation of the applied bias voltage from 0.05 V to 0.95 V induced a variation of about 10 % in the anisotropy. These voltages correspond to an electric field range from 0.085 to 1.6 GV/m (with $\epsilon = 3.4$ for the MgO barrier). In contrast to previous studies [74, 75] and the results of the MOKE measurements, the out-of-plane anisotropy increases with a positive voltage applied to the top electrode. Ni film thicknesses of 7 ML and 13 ML show smaller MEC but with the same tendency of increasing out-of-plane anisotropy with increasing electric field. Similar to the observations in the MOKE measurements, MEC is strongest for the sample with a thickness close to the SRT (here 11 ML).

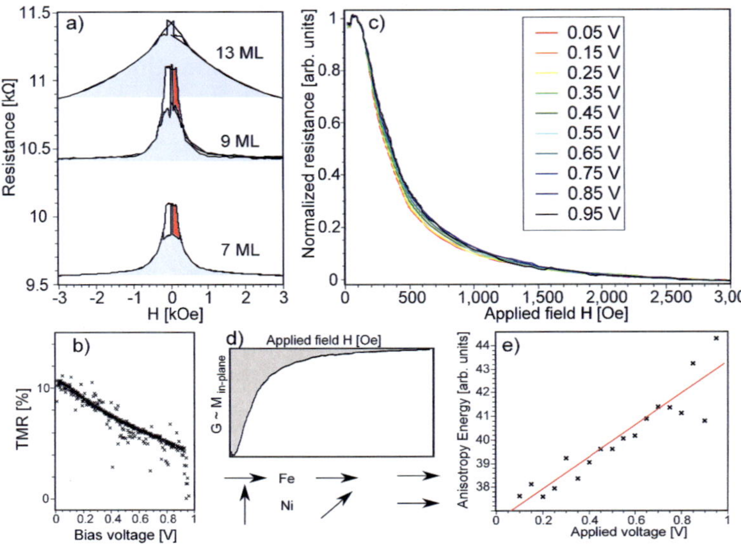

Figure 6.8: a) Resistance measured as a function of magnetic field. A 13 ML thick Ni film displays a continuous but slow decrease of the resistance with increasing field, indicative of a dominant out-of-plane anisotropy. Ni films with thicknesses of 9 ML and 7 ML show a small plateau at low magnetic fields and an abrupt decrease of the resistance with increasing magnetic fields, indicative of an in-plane easy axis of the magnetization at low magnetic fields. For all three curves, the symmetric contribution is shown in gray, the antisymmetric part is shown in red. Note that, more precisely, it is the conductances that add up. b) TMR ratio as a function of applied voltage for the 11 ML thick Ni film. c) Normalized resistance versus magnetic field, measured at applied bias voltages varied from 0.05 V to 0.95 V for a 11 ML thick Ni layer. d) Same measurement as c); variation of the conductance as a function of the applied field in combination with a possible orientation of the magnetization directions. The gray area is proportional to the out-of-plane anisotropy energy. e) Relative change of the magnetic anisotropy energy as a function of applied voltage, calculated from the data presented in c).

6.3 STM measurements on Ni/Cu(100)

It was shown in chapter 4 that STM can be a suitable tool to study MEC. Therefore, we also attempted to trigger the SRT in Ni/Cu(100) by the electric field in the STM junction. To do so, Ni was deposited by MBE onto a Cu(100) single crystal which was cleaned as described above for the case of a Cu(111) surface (see paragraph 4). During the Ni deposition, the polar MOKE signal and its derivative were measured. The latter corresponds to the magnetic susceptibility. It diverges at the SRT and therefore it was used to find the critical thickness. The deposition of Ni was stopped when a peak in the signal appeared (see Fig. 6.9a)). The typical topography of a such prepared sample (nominal Ni film thickness is about 10 ML) is shown in Fig. 6.9b), and it is in agreement with previous STM studies [84, 80]. Several issues opposed successful STM experiments on MEC in this system: the difference between the two states before and after the SRT is purely magnetic and thus it can only be detected with SP-STM. The difference in the dI/dU signal for the two magnetic states has to be extracted at the same energy because at different energies (different applied bias = different electric field) the signal is different anyway. This is why SP-STM can only reveal an electric-field-induced phase transition in a bistable system: either the state is manipulated at high electric field but can be read out at low electric field, or domains of different magnetic order coexist and can be imaged at the same time. This is essentially equal to the necessity of a system that undergoes a phase transition of first order. The SRT in 7 ML Ni/Cu(100), however, is most probably a transition of second order [72]. Another issue is the small field of view of an STM, it is probably too small for the relatively large magnetic domains. Especially the acquisition of dI/dU maps for SP-STM requires some time: in order to obtain a reasonable fraction of measured pixels that are not influenced by features of the step edges, the number of pixels should not be too low. But each pixel requires at least an integration time of about $1\,\mathrm{ms}$. Practically, this limits the dI/dU maps to about $1\,\mu\mathrm{m}^2$. (If we want to acquire less than $10\,\%$ of the measurement points on steps and if we have an average terrace size of $10\,\mathrm{nm}$, we will need a resolution of 1024×1024 pixels on $1\,\mu\mathrm{m}^2$, which will take $35\,\mathrm{min}$.) The size of the domains, however, is expected to be several micrometers, separated by domain walls of several $100\,\mathrm{nm}$ [85, 70]. Finally, the preparation system that is connected to the STM chamber did not allow deposition of wedge-shaped films, therefore the critical thickness had to be exactly matched during deposition. As is indicated by the results obtained by MOKE and TMR measurements on this system, strong MEC is only expected very close to the critical thickness for the SRT. Moreover, in our STM measurements, two critical parameters were changed after deposition: the surface had to be mildly annealed in order to decrease the surface roughness (see Fig. 6.9b)), which is necessary for a reasonable dI/dU map on a large scale. The change in surface roughness and a possible intermixing at the Ni/Cu interface change the surface and interface anisotropy contributions and thus shift the critical thickness for the SRT [80]. Furthermore, the STM measurements were carried out at $5\,\mathrm{K}$, while film growth and susceptibility measurement were done at room temperature. This is a problem because the critical thickness for SRT is temperature-dependent (see Fig. 6.1). It is not even sure whether at very low temperatures an out-of-plane easy magnetization direction occurs in this system [71]. Although for each issue, considered separately, a workaround may exist, for a successful experiment, all problems had to be solved at the same time. Unfortunately, this was not possible in spite of several months of measurement time, and STM-triggered MEC in Ni/Cu(100) could not be observed.

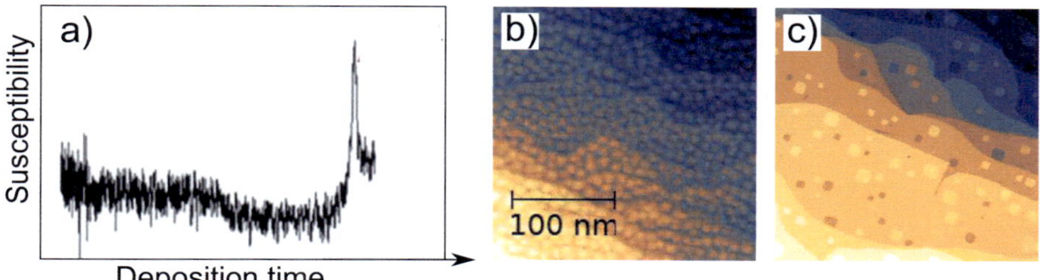

Figure 6.9: a) Magnetic susceptibility measured during deposition of Ni/Cu(100). b) Typical STM topography obtained on a Ni film deposited at room temperature. c) Topography of the annealed surface (same size as in b)) showing reduced surface roughness.

7 Discussion and conclusion

7.1 Structural relaxations of a metal surface in an electric field

Possible mechanisms that might explain the observed electric-field-induced modifications of the sample shall be discussed in this part. In the experiments on solid-state junctions presented in chapter 6, microscopic effects at the interface cannot be observed directly. Therefore, only general considerations can be made.

In the case of 2 ML Fe/Cu(111) and Fe/Ni(111), MEC is mediated by a crystallographic phase transition. As a consequence, the impact of electric fields on the lattice relaxation of a metallic surface in vacuum is important. In our STM measurements, at least lateral displacements within the surface can be followed atom by atom and a careful discussion of possible relaxation mechanisms is of interest. Although a rigorous separation is of course not possible, contributions to MEC by changes in electronic structure shall be neglected.

7.1.1 Response of an ideal infinite metal surface to a homogeneous electric field

In the following, the response of an infinite flat metal surface (in vacuum) on an electric field is derived in three steps, discussing the charge distribution, the electron potential and vertical atomic displacements.

E = 0

In the absence of any externally applied electric field, a dipole charge forms at the surface of a metal (see Fig. 7.1a)), which was first described by Smoluchowski [86]. The first theoretical explanation was given by Lang and Kohn [87] within density functional theory. They modeled the bulk as a combination of a homogeneous positively charged background and the mobile electrons, called the jellium model. At the surface, the sharp edge of the positively charged background cannot be fully balanced by the electrons which have finite wavelength and hence spill out into the vacuum. This imperfect balance of positive and negative charges leads to a dipole at the surface. The huge charge densities involved in this process (several electrons per nm^2) lead to considerable displacements of the atoms in the topmost layers. This relaxation of a few percent of the interlayer distance is most often directed inwards and can then be explained by the increased number of electrons contributing to bonding between surface and subsurface atoms [88]. But, depending on element and lattice direction, the relaxation may also exhibit an oscillatory behavior across several layers or may be directed outwards (see e.g. [89]). The dipole charge at the surface increases the potential barrier for the electrons at the surface by Φ_s. This adds up with the difference between the image potential E_I (due to the classical mirror charge) and the Fermi energy E_F to the work function $\Phi = E_I - E_F + \Phi_s$ (see Fig. 7.1a)).

E > 0, classical

In an electric field, additional charges are displaced in order to screen the bulk. This is the classical screening charge σ that is related to the electric field E by Gauss's theorem via the electric constant ϵ_0: $\sigma = \epsilon_0 E$ (see Fig. 7.1b)). Its centroid gives the position of the electric surface which is where the electric field starts and which is identical with the image plane. It is purely formed by the mobile electrons, and it is identical to the net charge of the considered surface. The number of charges involved is more than one order of magnitude lower than the one involved in the Smoluchowski effect described above, even in the case of the highest fields in our STM ($10\,\mathrm{GV/m} \approx 0.6$ electrons per nm^2). The total force related to this induced screening charge is given by the classical attractive force of two plates of a capacitor (i.e., between STM tip and sample) $F = \frac{1}{2}\epsilon_0 \cdot A \cdot E^2$, where A is the area of the electrodes. This leads to an expansion of the whole sample (see Fig. 7.1b), all atoms are displaced) and the tip. This might even lead to a contact of tip and sample [54]. The electric field outside the surface is reflected in the linear slope of the potential. A positive electric field leads to a reduction of the barrier height and a finite width. For high electric fields, this leads to a finite probability of the electrons to tunnel through the barrier (field electron emission).

E $>$ 0, semi-classical

It was first observed by Drechsler [90] with field ion microscopy that an external electric field also induces an additional dipole charge in the surface. When the lateral inhomogeneous distribution of positive charge (an array of ions) is taken into account, *ab-initio* calculations reproduce this observation. Fig. 7.1c) shows the induced charge density $\Delta\rho = \rho(E > 0) - \rho(E = 0)$ at a Ag(001) surface in an electric field [91, 92]. It can be seen that the 'monopole'-like classical screening charge is mainly located on top of the surface atomic layer, but that there is also electron accumulation in between and electron depletion underneath the ion cores, which leads to an additional 'dipole'-like induced charge. This redistribution of charges within the topmost atomic layer leads to further displacements of the atoms proportional to the electric field [93, 94]. This can be understood in a simple surface-chemical picture of filling of interatomic bonds: an increase of the electron density between the atoms leads to stronger bonds and a compressive interlayer stress. The sign of the resulting displacement corresponds to the displacement of a positive ion in the (imperfectly screened) external electric field [95]. This displacement is called *stretch*, in contrast to the vertical *strain* induced by the screening charge. The additional dipole further contributes to the work function.

7.1.2 Relaxations in 2 ML Fe/Cu(111) and 1 ML Fe/Ni(111) in the STM setup

The redistribution of electrons in the surface also affects the lateral interatomic bonds within the topmost atomic layer of the surface. An increase (decrease) of the electron density leads to stronger (weaker) bonds and results in a tensile (compressive) surface stress. The dependence of the surface stress as a function of the charge density $\frac{\partial \tau}{\partial \sigma}$, called surface stress-charge coefficient, can be derived in a simple approximation from the jellium model [15] and is then $\frac{\partial \tau}{\partial \sigma} = -0.13 \cdot r_s$, where r_s^3 is the volume per electron in atomic units. In general, experimental values are higher, e.g. Au(111) shows a value of $-1.9\,\mathrm{V}$ [93]. This effect is of little interest in the case of a large homogeneous planar capacitance. But in our case of an STM setup, we locally apply an electric field, which means that the sample is locally strained in-plane. Moreover, the sample structure is also laterally inhomogeneous.

The possible influence of vertical and lateral forces and relaxations on the two experimentally studied systems (see chapter 4, 5) are discussed in the following.

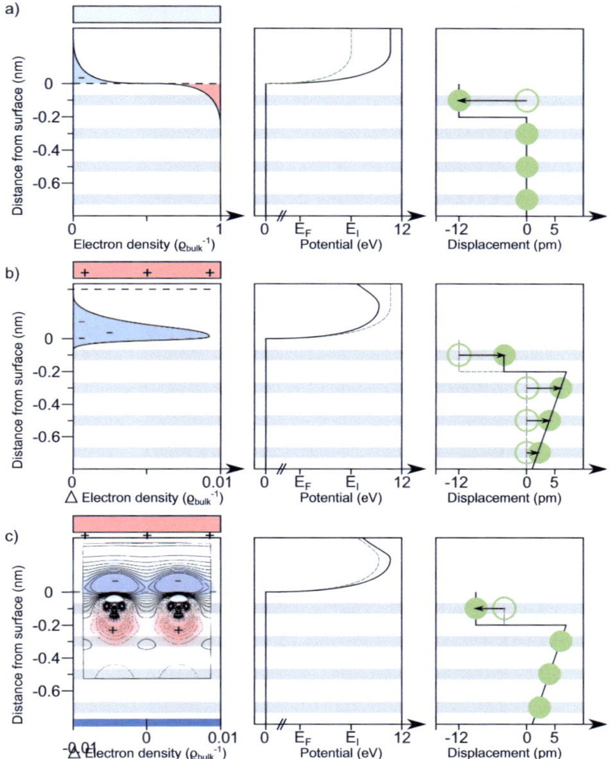

Figure 7.1: Metal surface (bottom electrode) in the homogeneous electric field of a thin (charged) plate (top electrode), separated by a vacuum barrier. The gray horizontal bars give the positions of the ion cores. a) $E = 0$, jellium model: the finite wavelength of the electrons leads to a surface dipole (e^- excess blue, deficiency red). This is reflected in an increase of the potential barrier for the electrons and a considerable displacement of the ion cores depicted in green (dashed lines describe the situation before). b) $E > 0$, classical: the screening charge (here: negative, blue) is superpositioned with the surface dipole from a) (not shown for simplicity). The metal surface is attracted by the electrode, which leads to an outward displacement of the atomic cores. c) $E > 0$, discrete lateral distribution of positive ions: besides the classical screening charge above the topmost atomic layer (blue), also in between the atoms charges are redistributed [91]. This additional dipole-like charge adds up to the work function and induces further displacements of the atoms.

1 ML Fe/Ni(111)

In the case of 1 ML Fe/Ni(111), we found a coexistence of fcc and hcp stacking. The surface planes of these two lattices are identical and the difference of the interlayer distances is very small. Therefore, the energy balance between these two states is not expected to be changed by application of (vertical or lateral) strain. Fig. 5.6, however, clearly shows an enlarged hcp domain for higher electric fields. This can be explained by lateral forces on the domain boundary that result from the inhomogeneous surface stress. Fig. 7.2a) shows a schematic of a light fcc-hcp domain wall in the presence of the tip. The lateral gradient of the electric field induces a lateral gradient of surface charge which in turn gives a gradient of the interatomic bonding strength. This results in a position-dependent domain wall energy or, in other words, in a lateral force on the domain wall. Depending on the polarity of the electric field, the tip attracts or repells the domain wall. This mechansim might be supported by a reduction of the barrier between fcc and hcp positions by an increase of the interlayer distance due to electrostatic attraction between tip and sample.

2 ML Fe/Cu(111)

The situation is more complicate for 2 ML Fe/Cu(111): the coexisting fcc and bcc stackings have different in-plane lattice constants and different interlayer distances (see Fig. 4.2). This is why both the vertical and the lateral relaxations change the difference between the energy levels of the fcc and the bcc state: an increase of electron density (positive STM tip) in the surface leads to stronger (vertical and lateral) bonding and thus to a reduced unit cell volume. This favors the fcc lattice, while the bcc lattice is favored by a reduced electron density (negative STM tip). These mechanisms are summarized in Fig. 7.2b) and lead to the following interpretation of the experimental results on electric-field-dependent switching dynamics obtained for a fixed position of the tip (see paragraph 4.4.6): the polarity-independent increase of the switching frequency (i.e., the decrease of the barrier between fcc and bcc) can be identified with the classical electrostatic attraction which scales with the square of the electric field (black solid line in Fig. 7.2b)). This is in agreement with the *ab-intio*-calculated two-dimensional energy landscape (see Fig. 4.7c)). The polarity-dependent stabilization of the fcc (bcc) lattice at high values of the electric field is explained by the increase (decrease) of surface strain (red line in Fig. 7.2b)). Similarly, in the electric-field-free case, it is the local variation of the in-plane strain that leads to the coexistence of fcc and bcc lattices in the islands. The difference between the work functions of tip and sample and the particular value of the stretch-charge coefficient (see 7.1.1) may shift the apex of the parabola to different values of the applied bias voltage. The influence of an electric field (surface charge) is of course not the same for fcc (111) and for bcc (110) surfaces, and the strain-electric-field dependence sketched in in Fig. 7.2b) is probably different for these two lattices. On top of this, the mechanism that induces a lateral force on the domain boundary (which was found to be a heavy one, see Fig. 4.1) also holds for the fcc-bcc boundary. Furthermore, the lateral misfit strain within an island is already inhomogeneous in the electric-field-free case, and it depends on the particular distribution of fcc and bcc domains. This is why it is difficult to predict the preferred switching direction for a particular configuration of the STM tip close to a domain boundary (see Fig. 7.2b)). The shape of the tip apex, for example, is crucial for the gradient of the electric field on the surface.

In summary, *ab-intio* calculations that are carefully adapted to the particular system are necessary to fully understand the microscopic processes that lead to the electric-field-induced phase transitions in 2 ML Fe/Cu(111) and 1 ML Fe/Ni(111).

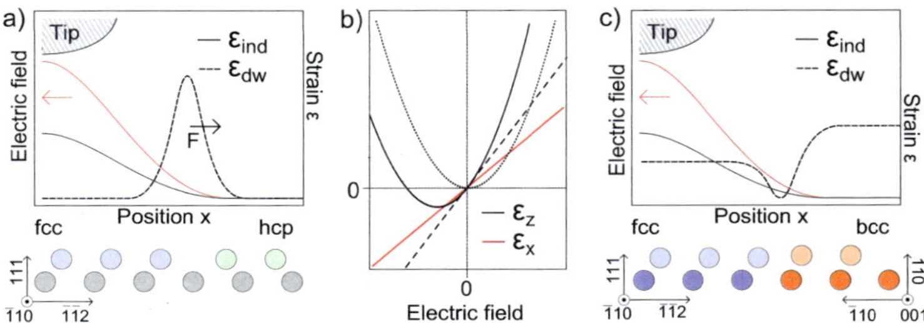

Figure 7.2: a) One-dimensional model of the fcc-hcp domain boundary in 1 ML Fe/Ni(111) in the electric field of an STM tip: the sphere model shows a $(\bar{1}10)$ plane across the domain boundary between the fcc (blue) and hcp (green) Fe lattice on Ni (gray). The strain variation ϵ_{dw} due to the misfit dislocation is depicted with the dashed black line. The lateral variation of the intensity of the electric field (perpendicular to the surface) of the STM tip is drawn in red. It results in a lateral variation of the in-plane strain ϵ_{ind} (solid black line). A net force F on the domain wall results. b) Qualitative electric field dependence of lattice relaxations of the topmost atomic cores in a metal surface: the vertical displacement due to the electrostatic attraction scales with E^2 (black dotted line), while the stretch (black dashed line) scales linearly with E. The total vertical displacement, ϵ_z, is shown by the black solid line. The value of the lateral strain ϵ_x is shown by the red line. c) Similar to a), for the case of a boundary between fcc (blue) and bcc (orange) domains in 2 ML Fe/Cu(111). Here the bcc lattice is shown as an expanded fcc lattice (which is energetically not correct).

7.2 Conclusion

In this work, we have shown that STM is an ideal tool to study MEC because it combines surface analysis with the possibility to apply high electric fields locally. By using STM, we brought the first proof of principle of writing nanometer-sized magnetic bits with an electric field. Three major differences of our STM experiments (2 ML Fe/Cu(111)) compared to previous studies on MEC in metals are to be emphasized: a metal surface in vacuum is the simplest possible interface, which allows a thorough understanding of the microscopic processes that lead to MEC. Furthermore, we succeeded in a bistable switching of the *magnetic order* of the sample, not of the *magnetization direction*. This is particularly interesting with regard to possible applications in future memory devices. Finally, STM is the only technique that allows to study MEC on the nanoscale. The observed effect is not a peculiarity of Fe because similar structural and magnetic phase transitions occur in a whole variety of transition metals. In addition to the structurally-mediated MEC observed in 2 ML Fe/Cu(111), purely magnetic effects like SRTs due to changes of the surface anisotropy are promising candidates for spin-polarized STM studies. We are optimistic that STM will further advance the exciting development of MEC at metal surfaces.

Deutsche Zusammenfassung: Magnetoelektrische Kopplung an Metalloberflächen

Grundlagenforschung in der Physik bedarf prinzipiell keiner Begründung durch mögliche technische Anwendungen. Im Fall dieser Arbeit aber liegt die mögliche Verwendung in zukünftigen Speichermedien auf der Hand und wird deshalb im Folgenden genannt, um die Experimente über magnetoelektrische Kopplung an Metalloberflächen zu motivieren. In heutzutage üblichen Festplatten wird die binäre Information in Form von Magnetisierungsrichtungen mikroskopisch kleiner magnetischer Einheiten abgespeichert. Um neue Informationen abzulegen, wird die Magnetisierung dieser Bits durch ein angelegtes Magnetfeld einer Spule ausgerichtet. Die bereits seit vielen Jahren fortwährende Entwicklung von Computerfestplatten mit immer höheren Speicherdichten geht einher mit einer Verkleinerung der einzelnen Bits. Während es lange Zeit eine eher technische Herausforderung war, sämtliche Bauteile zu verkleinern, werden mittlerweile grundlegende physikalische Grenzen erreicht. Die thermische Stabilität der Magnetisierungsrichtung eines einzelnen Bits, also die Lebensdauer der gespeicherten Information, hängt von der magnetischen Anisotropieenergie, d. h. dem Produkt aus Volumen und magnetischer Anisotropiekonstante, ab. Verringert man nun das Volumen, muss im Gegenzug ein Material höherer Anisotropie verwendet werden, um die Lebensdauer aufrechtzuerhalten. Dann werden aber auch für die Magnetisierungsänderung beim Schreiben von Informationen höhere Magnetfelder benötigt. Die

Magnetfelder, die in den Schreibköpfen aktueller Festplatten verwendet werden, sind aber bereits an der Grenze des Möglichen. Deshalb wird mit dieser Technik eine weitere Erhöhung der Speicherdichte zwangsläufig mit einer verringerten Lebensdauer der gespeicherten Information erkauft. Einen Ausweg aus diesem Dilemma bietet die Möglichkeit, magnetische Eigenschaften mit Hilfe der magnetoelektrischen Kopplung durch ein elektrisches Feld zu verändern. Weitere Vorteile wären ein geringerer Energieverbrauch und eine höhere mögliche Schreibgeschwindigkeit, beides aufgrund der dann nicht mehr benötigten Spule zur Magnetfelderzeugung. Magnetoelektrische Kopplung wurde bisher vor allem in komplizierten Oxidstrukturen beobachtet, Metalle scheinen auf den ersten Blick ungeeignet: Zu ihnen gehören zwar die typischen ferromagnetischen Materialien, ein äußeres elektrisches Feld wird aber von der Oberflächenladung des Metalls vollständig abgeschirmt. Allerdings erstreckt sich diese Oberflächenladung bis in die äußerste Atomlage, je nach Stärke des elektrischen Felds mehr oder weniger weit, und führt dort zu kleinen Positionsveränderungen der Atomrümpfe. Findet man also ein System, in dem der Einfluss der Oberfläche auf das gesamte System nicht vernachlässigbar ist, wie es zum Beispiel in ultradünnen Filmen der Fall ist, dann ergibt sich auf diese Weise eine Möglichkeit, die Kristallstruktur eines Metalls durch ein elektrisches Feld gezielt zu verändern [9]. Magnetoelektrische Kopplung tritt auf, wenn diese Änderung der Kristallstruktur auch eine Änderung der magnetischen Ordnung hervorruft [11]. In der vorliegenden Arbeit wurde dieser Effekt an drei verschiedenen Systemen und mit unterschiedlichen Methoden untersucht. Es ist bekannt, dass die magnetische Ordnung in Eisen eng mit der Kristallstruktur verknüpft ist und empfindlich von kleinen Änderungen der Packungsdichte abhängen kann. Deshalb bietet sich Eisen als Modellsystem an. Für die Untersuchung der magnetoelektrischen Kopplung auf der Nanometerskala ist ein Rastertunnelmikroskop (RTM) ideal geeignet, da es bekanntermaßen eine extrem hohe räumliche Auflö-

114

sung bietet, aber auch die nötigen, sehr hohen elektrischen Felder im Bereich zwischen Spitze und Probe erzeugen kann. Zuerst wurden zweilagige Eiseninseln auf einem Kupfersubstrat untersucht. Diese Eiseninseln zeigen zwei unterschiedliche Bereiche: eine antiferromagnetische kubisch-flächenzentrierte Phase im Zentrum und eine ferromagnetische kubisch-raumzentrierte Phase im äußeren Bereich der Inseln. Es wurde gezeigt, dass mit Hilfe des elektrischen Felds der RTM-Spitze (etwa $1\,\mathrm{GV/m}$) kleine Bereiche von etwa $2\,\mathrm{nm}^2$ zwischen diesen beiden Phasen reproduzierbar und deterministisch hin- und hergeschaltet werden können. Dieser induzierte Phasenübergang konnte mit Hilfe von Ab-initio-Rechnungen [38] dadurch erklärt werden, dass die vergrößerten interatomaren Abstände bei einer positiven Oberflächenladung den ferromagnetischen kubischraumzentrierten Zustand energetisch bevorzugen, während eine negative Oberflächenladung zum antiferromagnetischen kubischflächenzentrierten Zustand führt. Anhand eines Modells einer zweidimensionalen Energielandschaft wurde das dynamische Verhalten dieses Phasenübergangs in Abhängigkeit von der Feldstärke und der mechanischen Verspannung der Eisendoppellage untersucht. Im Rahmen dieser Arbeit ist es erstmals gelungen, die magnetische Ordnung auf der Nanometerskala durch ein elektrisches Feld zwischen zwei stabilen Konfigurationen hin- und herzuschalten [38]. Das zweite untersuchte System ist eine ausgedehnte Eisenmonolage auf einem Nickelsubstrat. Im RTM konnte mit atomarer Auflösung gezeigt werden, dass sich die Gitterstruktur dieses Eisenfilms zwischen der kubisch-flächenzentrierten Ordnung des Substrats und einer, ebenfalls stabilen, hexagonal dichtestgepackten Ordnung mit Hilfe des elektrischen Felds hin- und herschalten lässt. Dabei ändert sich auch die magnetische Ordnung geringfügig. Leider lässt sich ein Vakuum-Tunnelkontakt wie im RTM in technischen Anwendungen nur sehr schwer realisieren. Deshalb wurde als drittes System ein Nickelfilm auf einem Kupfersubstrat gewählt, an den über eine Magnesiumoxidschicht ein elektrisches Feld angelegt wurde. Ein solcher

Festkörpertunnelkontakt könnte in bestehende Halbleitertechniken integriert weden. Durch Messungen mit Hilfe des magnetooptischen Kerr-Effekts und des Tunnelmagnetowiderstands konnte gezeigt werden, dass dieses elektrische Feld die Magnetisierung im Nickelfilm reproduzierbar verändert. Eine Vielzahl von ähnlichen Phasenübergängen in dünnen Metallfilmen bietet zahlreiche weitere Möglichkeiten für magnetoelektrische Kopplung, sei es indirekt über die Veränderung des Kristallgitters oder direkt über eine Veränderung der elektronischen Eigenschaften der Oberfläche. Die jüngsten Fortschritte auf diesem Gebiet [13] lassen hoffen, dass eines Tages tatsächlich ultraschnelle und energiesparende Massenspeicher entwickelt werden, die auf dem Prinzip der magnetoelektrischen Kopplung basieren.

Bibliography

[1] J. S. Moodera, L. R. Kinder, T. M. Wong, and R. Meservey. "Large magnetoresistance at room temperature in ferromagnetic thin film tunnel junctions". In: *Physical Review Letters* 74 (1995), pp. 3273–3276 (cit. on p. 1).

[2] S. Yuasa, T. Nagahama, A. Fukushima, Y. Suzuki, and K. Ando. "Giant room-temperature magnetoresistance in single-crystal Fe/MgO/Fe magnetic tunnel junctions." In: *Nature Materials* 3 (2004), pp. 868–871 (cit. on pp. 1, 93).

[3] J. Mallinson. "A new theory of recording media noise". In: *IEEE Transactions on Magnetics* 27 (1991), pp. 3519–3531 (cit. on p. 1).

[4] T. Lottermoser, T. Lonkai, U. Amann, D. Hohlwein, J. Ihringer, and M. Fiebig. "Magnetic phase control by an electric field". In: *Nature* 430 (2004), pp. 541–544 (cit. on p. 3).

[5] H. Zheng, J. Wang, S. E. Lofland, Z. Ma, L. Mohaddes-Ardabili, T. Zhao, L. Salamanca-Riba, S. R. Shinde, S. B. Ogale, F. Bai, D. Viehland, Y. Jia, D. G. Schlom, M. Wuttig, A. Roytburd, and R. Ramesh. "Multiferroic $BaTiO_3$-$CoFe_2O_4$ nanostructures". In: *Science* 303 (2004), pp. 661–663 (cit. on pp. 3, 25).

[6] N. A. Spaldin and M. Fiebig. "The renaissance of magneto-electric multiferroics". In: *Science* 309 (2005), pp. 391–392 (cit. on pp. 3, 24, 25).

[7] M. Fechner, I. V. Maznichenko, S. Ostanin, A. Ernst, J. Henk, P. Bruno, and I. Mertig. "Magnetic phase transition in two-phase multiferroics predicted from first principles". In: *Physical Review B* 78 (2008), p. 212406 (cit. on pp. 3, 25).

[8] C.-G. Duan, S. S. Jaswal, and E. Y. Tsymbal. "Predicted magnetoelectric effect in $Fe/BaTiO_3$ multilayers: ferroelectric control of magnetism". In: *Physical Review Letters* 97 (2006), p. 047201 (cit. on pp. 3, 25).

[9] J. Weissmüller, R. N. Viswanath, D. Kramer, P. Zimmer, R. Würschum, and H. Gleiter. "Charge-Induced Reversible Strain in a Metal". In: *Science* 300 (2003), pp. 312–315 (cit. on pp. 3, 26, 114).

[10] S. Blügel and X. Nie. "Electric field for magnetic reversal of a thin film". European Patent. (2000) (cit. on pp. 3, 26).

[11] M. Weisheit, S. Fähler, A. Marty, Y. Souche, C. Poinsignon, and D. Givord. "Electric field-induced modification of magnetism in thin-film ferromagnets." In: *Science* 315 (2007), pp. 349–51 (cit. on pp. 3, 26, 114).

[12] T. Maruyama, Y. Shiota, T. Nozaki, K. Ohta, N. Toda, M. Mizuguchi, A. A. Tulapurkar, T. Shinjo, M. Shiraishi, S. Mizukami, Y. Ando, and Y. Suzuki. In: *Nature Nanotechnology* 4 (2009), p. 158 (cit. on pp. 4, 26).

[13] Y. Shiota, T. Nozaki, F. Bonell, S. Murakami, T. Shinjo, and Y. Suzuki. "Induction of coherent magnetization switching in a few atomic layers of FeCo using voltage pulses." In: *Nature Materials* 11 (2012), pp. 39–43 (cit. on pp. 4, 26, 92, 116).

[14] V. L. Moruzzi, P. M. Marcus, K. Schwarz, and P. Mohn. "Ferromagnetic phases of bcc and fcc Fe, Co, and Ni". In: *Physical Review B* 34 (1986), pp. 1784–1791 (cit. on pp. 4, 37).

[15] H. Ibach. *Physics of Surfaces and Interfaces.* Springer Berlin Heidelberg, (2006), p. 652 (cit. on pp. 8, 56, 106).

[16] M. T. Johnson, P. J. H. Bloemen, F. J. A. den Broeder, and J. J. de Vries. "Magnetic anisotropy in metallic multilayers". In: *Reports on Progress in Physics* 59 (1996), p. 1409 (cit. on p. 10).

[17] M. Jullière. "Tunneling between ferromagnetic films". In: *Physics Letters A* 54 (1975), pp. 225–226 (cit. on p. 11).

[18] J. C. Slonczewski. "Conductance and exchange coupling of two ferromagnets separated by a tunneling barrier". In: *Physical Review B* 39 (1989), pp. 6995–7002 (cit. on pp. 12, 50).

[19] R. Young, J. Ward, and F. Scire. "The Topografiner: an instrument for measuring surface microtopography". In: *Review of Scientific Instruments* 43 (1972), pp. 999–1011 (cit. on p. 14).

[20] G. Binnig, H. Rohrer, C. Gerber, and E. Weibel. "Tunneling through a controllable vacuum gap". In: *Applied Physics Letters* 40 (1982), pp. 178–180 (cit. on p. 14).

[21] G. Binnig, H. Rohrer, C. Gerber, and E. Weibel. "Surface studies by scanning tunneling microscopy". In: *Physical Review Letters* 49 (1982), pp. 57–61 (cit. on p. 14).

[22] J. Bardeen. "Tunnelling from a many-particle point of view". In: *Physical Review Letters* 6 (1961), pp. 57–59 (cit. on p. 15).

[23] J. Tersoff and D. R. Hamann. "Theory of the scanning tunneling microscope". In: *Physical Review B* 31 (1985), pp. 805–813 (cit. on p. 16).

[24] V. A. Ukraintsev. "Data evaluation technique for electron-tunneling spectroscopy". In: *Physical Review B* 53 (1996), pp. 11176–11185 (cit. on p. 17).

[25] B. J. van Wees, H. van Houten, C. W. J. Beenakker, J. G. Williamson, L. P. Kouwenhoven, D. van der Marel, and C. T. Foxon. "Quantized conductance of point contacts in a two-dimensional electron gas". In: *Physical Review Letters* 60 (1988), pp. 848–850 (cit. on p. 20).

[26] W. Eerenstein, N. D. Mathur, and J. F. Scott. "Multiferroic and magnetoelectric materials". In: *Nature* 442 (2006), pp. 759–765 (cit. on p. 24).

[27] W. Röntgen. In: *Annalen der Physik und Chemie* 35 (1888), p. 264 (cit. on p. 24).

[28] P. Curie. "Sur la symétrie dans les phénomènes physiques, symétrie d'un champ électrique et d'un champ magnétique". In: *Journal de Physique Théorique et Appliquée* 3 (1894), pp. 393–415 (cit. on p. 24).

[29] D. N. Astrov. In: *Soviet physics, JETP* 11 (1960), p. 708 (cit. on p. 25).

[30] D. N. Astrov. In: *Soviet physics, JETP* 13 (1961), p. 729 (cit. on p. 25).

[31] I. E. Dzyaloshinskii. In: *Soviet physics, JETP* 10 (1960), p. 628 (cit. on p. 25).

[32] S. Dong and J.-M. Liu. "Recent progress of multiferroic perovskite manganites". In: *Modern Physics Letters B* 26 (2012), p. 1230004 (cit. on p. 25).

[33] N. A. Hill. "Density functional studies of multiferroic magnetoelectrics". In: *Annual Review of Materials Research* 32 (2002), pp. 1–37 (cit. on p. 25).

[34] F. Zavaliche, H. Zheng, L. Mohaddes-Ardabili, S. Yang, Q. Zhan, P. Shafer, E. Reilly, R. Chopdekar, Y. Jia, P. Wright, D. Schlom, Y. Suzuki, and R. Ramesh. "Electric field-induced magnetization switching in epitaxial columnar nanostructures". In: *Nano Letters* 5 (2005), pp. 1793–1796 (cit. on p. 25).

[35] H. Ohno, D. Chiba, F. Matsukura, T. Omiya, E. Abe, T. Dietl, and Y. Ohno. "Electric-field control of ferromagnetism". In: *Nature* 408 (2000), p. 944 (cit. on p. 25).

[36] T. Balashov. *Inelastic scanning tunneling spectroscopy: magnetic excitations on the nanoscale*. Dissertation, Universität Karlsruhe. (2009) (cit. on p. 31).

[37] L. Gerhard. *Magnetoelektrischer Phasenübergang in Fe/Cu(111)*. Diplomarbeit, Universität Karlsruhe. (2008) (cit. on p. 37).

[38] L. Gerhard, T. K. Yamada, T. Balashov, A. F. Takacs, R. J. H. Wesselink, M. Däne, M. Fechner, S. Ostanin, A. Ernst, I. Mertig, and W. Wulfhekel. "Magnetoelectric coupling at metal surfaces". In: *Nature Nanotechnology* 5 (2010), pp. 792–797 (cit. on pp. 37, 41, 43, 46, 49, 115).

[39] L. Gerhard, T. K. Yamada, T. Balashov, A. F. Takacs, R. J. H. Wesselink, M. Däne, M. Fechner, S. Ostanin, A. Ernst, I. Mertig, and W. Wulfhekel. "Electrical control of the magnetic state of Fe". In: *IEEE Transactions on Magnetics* 47 (2011), pp. 1619–1622 (cit. on p. 37).

[40] T. K. Yamada, L. Gerhard, T. Balashov, A. F. Takacs, R. J. H. Wesselink, and W. Wulfhekel. "Electric field control of Fe nano magnets: towards metallic nonvolatile data storage devices". In: *Japanese Journal of Applied Physics* 50 (2011) (cit. on pp. 37, 47).

[41] A. Biedermann, M. Schmid, and P. Varga. "Nucleation of bcc iron in ultrathin fcc films". In: *Physical Review Letters* 86 (2001), pp. 464–467 (cit. on p. 37).

[42] A. Biedermann, W. Rupp, M. Schmid, and P. Varga. "Coexistence of fcc- and bcc-like crystal structures in ultrathin Fe films grown on Cu(111)". In: *Physical Review B* 73 (2006), p. 165418 (cit. on pp. 37–39, 41).

[43] T. Michely, M. Hohage, M. Bott, and G. Comsa. "Inversion of growth speed anisotropy in two dimensions". In: *Physical Review Letters* 70 (1993), pp. 3943–3946 (cit. on p. 38).

[44] E. C. Bain. "The nature of martensite". In: *Transactions of the A.I.M.E.* 70 (1924), pp. 25–46 (cit. on p. 39).

[45] G. Kurdjumov and G. Sachs. "Über den Mechanismus der Stahlhärtung". In: *Zeitschrift für Physik* 64 (1930), pp. 325–343 (cit. on p. 39).

[46] L. Sandoval, H. M. Urbassek, and P. Entel. "The Bain versus Nishiyama-Wassermann path in the martensitic transformation of Fe". In: *New Journal of Physics* 11 (2009), p. 103027 (cit. on p. 39).

[47] H. F. Berger and K. D. Rendulic. "Nozzle beam experiments on the adsorption system hydrogen/iron". In: *Surface Science* 251/252 (1991), p. 882 (cit. on p. 41).

[48] J. Shen, M. Klaua, P. Ohresser, H. Jenniches, J. Barthel, C. V. Mohan, and J. Kirschner. "Structural and magnetic phase transitions of Fe on stepped Cu(111)". In: *Physical Review B* 56 (1997), pp. 11134–11143 (cit. on p. 43).

[49] J. Shen, P. Ohresser, C. V. Mohan, M. Klaua, J. Barthel, and J. Kirschner. "Magnetic moment of fcc Fe(111) ultrathin films by ultrafast deposition on Cu(111)". In: *Physical Review Letters* 80 (1998), pp. 1980–1983 (cit. on p. 43).

[50] M. A. Torija, Z. Gai, N. Myoung, E. W. Plummer, and J. Shen. "Frozen low-spin interface in ultrathin Fe films on Cu(111)". In: *Physical Review Letters* 95 (2005), p. 027201 (cit. on p. 43).

[51] M. Lüders, A. Ernst, W. M. Temmerman, Z. Szotek, and P. J. Durham. "Ab initio angle-resolved photoemission in multiple-scattering formulation". In: *Journal of Physics: Condensed Matter* 13 (2001), p. 8587 (cit. on p. 44).

[52] L. Szunyogh, B. Újfalussy, P. Weinberger, and J. Kollár. "Self-consistent localized KKR scheme for surfaces and interfaces". In: *Physical Review B* 49 (1994), pp. 2721–2729 (cit. on p. 44).

[53] F. J. Jedema, A. T. Filip, and B. J. van Wees. "Electrical spin injection and accumulation at room temperature in an all-metal mesoscopic spin valve." In: *Nature* 410 (2001), pp. 345–348 (cit. on p. 50).

[54] W. A. Hofer, A. J. Fisher, R. A. Wolkow, and P. Grütter. "Surface relaxations, current enhancements, and absolute distances in high resolution scanning tunneling microscopy". In: *Physical Review Letters* 87 (2001), p. 236104 (cit. on pp. 50, 105).

[55] G. Kresse and J. Furthmüller. "Efficiency of ab-initio total energy calculations for metals and semiconductors using a plane-wave basis set". In: *Computational Materials Science* 6 (1996), pp. 15–50 (cit. on p. 50).

[56] J. Hafner. "Ab-initio simulations of materials using VASP: Density-functional theory and beyond". In: *Journal of Computational Chemistry* 29.13 (2008), pp. 2044–2078 (cit. on p. 50).

[57] R. C. Longo, E. Martinez, O. Dieguez, A. Vega, and L. J. Gallego. "Morphology and magnetism of Fe monolayers and small Fe n clusters (n = 2-19) supported on the Ni(111) surface". In: *Nanotechnology* 18 (2007), p. 055701 (cit. on p. 69).

[58] G. C. Gazzadi, F. Bruno, R. Capelli, L. Pasquali, and S. Nannarone. "Structural transition in Fe ultrathin epitaxial films grown on Ni(111)". In: *Physical Review B* 65 (2002), p. 205417 (cit. on p. 69).

[59] A. Theobald, V. Fernandez, O. Schaff, P. Hofmann, K.-M. Schindler, V. Fritzsche, A. M. Bradshaw, and D. P. Woodruff. "Structure of adsorbed Fe on Ni111". In: *Physical Review B* 58 (1998), pp. 6768–6771 (cit. on p. 69).

[60] R. Wu and A. J. Freeman. "Structural and magnetic properties of Fe/Ni(111)". In: *Physical Review B* 45 (1992), pp. 7205–7210 (cit. on p. 69).

[61] B. An, L. Zhang, S. Fukuyama, and K. Yokogawa. "Hydrogen adsorption on Fe monolayer grown on Ni(111) investigated by scanning tunneling microscopy". In: *Japanese Journal of Applied Physics* 46 (2007), pp. 5586–5590 (cit. on p. 70).

[62] B. An, L. Zhang, S. Fukuyama, and K. Yokogawa. "Growth and structural transition of Fe ultrathin films on Ni(111) investigated by LEED and STM". In: *Physical Review B* 79 (2009), pp. 1–7 (cit. on pp. 71–73).

[63] J. A. Stroscio, D. T. Pierce, A. Davies, R. J. Celotta, and M. Weinert. "Tunneling spectroscopy of bcc(001) surface states". In: *Physical Review Letters* 75 (1995), pp. 2960–2963 (cit. on p. 72).

[64] A. Theobald, V. Fernandez, O. Schaff, P. Hofmann, K.-M. Schindler, V. Fritzsche, A. M. Bradshaw, and D. P. Woodruff. "Structure of adsorbed Fe on Ni(111)". In: *Physical Review B* 58 (1998), pp. 6768–6771 (cit. on p. 72).

[65] B. Schulz and K. Baberschke. "Crossover from in-plane to perpendicular magnetization in ultrathin Ni/Cu(001) films". In: *Physical Review B* 50 (1994), pp. 13467–13471 (cit. on pp. 81, 86).

[66] W. L. O'Brien and B. P. Tonner. "Transition to the perpendicular easy axis of magnetization in Ni ultrathin films found by x-ray magnetic circular dichroism". In: *Physical Review B* 49 (1994), pp. 15370–15373 (cit. on p. 81).

[67] R. Vollmer, T. Gutjahr-Löser, J. Kirschner, S. van Dijken, and B. Poelsema. "Spin-reorientation transition in Ni films on Cu(001): the influence of H_2 adsorption". In: *Physical Review B* 60 (1999), pp. 6277–6280 (cit. on pp. 81, 86).

[68] W. L. O'Brien, T. Droubay, and B. P. Tonner. "Transitions in the direction of magnetism in Ni/Cu(001) ultrathin films and the effects of capping layers". In: *Physical Review B* 54 (1996), pp. 9297–9303 (cit. on p. 81).

[69] P. Poulopoulos and K. Baberschke. "Magnetism in thin films". In: *Journal of Physics: Condensed Matter* 11 (1999), p. 9495 (cit. on pp. 82, 83, 86).

[70] R. Ramchal. *In situ magnetic domain imaging at the spin-reorientation transition of ultrathin Ni- and Fe/Ni-Films*. Dissertation, Universität Duisburg-Essen. (2004) (cit. on pp. 82, 100).

[71] M. Farle, W. Platow, A. N. Anisimov, P. Poulopoulos, and K. Baberschke. "Anomalous reorientation phase transition of the magnetization in fct Ni/Cu(001)". In: *Physical Review B* 56 (1997), pp. 5100–5103 (cit. on pp. 82, 100).

[72] C. Klein, R. Ramchal, A. K. Schmid, and M. Farle. "Controlling the kinetic order of spin-reorientation transitions in Ni/Cu(100) films by tuning the substrate step structure". In: *Physical Review B* 75 (2007), p. 193405 (cit. on pp. 82, 99).

[73] D. Sander, W. Pan, S. Ouazi, J. Kirschner, W. Meyer, M. Krause, S. Müller, L. Hammer, and K. Heinz. "Reversible H-induced switching of the magnetic easy axis in Ni/Cu(001) thin films". In: *Physical Review Letters* 93 (2004), pp. 8–11 (cit. on p. 82).

[74] Y. Shiota, T. Maruyama, T. Nozaki, T. Shinjo, M. Shiraishi, and Y. Suzuki. "Voltage-Assisted Magnetization Switching in Ultrathin $Fe_{80}Co_{20}$ Alloy Layers". In: *Applied Physics Express* 2.6 (2009), p. 063001 (cit. on pp. 82, 83, 97).

[75] F. Bonell, S. Murakami, Y. Shiota, T. Nozaki, T. Shinjo, and Y. Suzuki. "Large change in perpendicular magnetic anisotropy induced by an electric field in FePd ultrathin films". In: *Applied Physics Letters* 98 (2011), p. 232510 (cit. on pp. 82, 83, 97).

[76] T. Nozaki, Y. Shiota, M. Shiraishi, T. Shinjo, and Y. Suzuki. "Voltage-induced perpendicular magnetic anisotropy change in magnetic tunnel junctions". In: *Applied Physics Letters* 96 (2010), p. 022506 (cit. on pp. 82, 92).

[77] W. Platow, U. Bovensiepen, P. Poulopoulos, M. Farle, K. Baberschke, L. Hammer, S. Walter, S. Müller, and K. Heinz. "Structure of ultrathin Ni/Cu(001) films as a function of film thickness, temperature, and magnetic order". In: *Physical Review B* 59 (1999), pp. 12641–12646 (cit. on p. 84).

[78] C. Lin, Y. H. Xu, H. Naramoto, P. Wei, S. Kitazawa, and K. Narumi. "Morphology evolution of thin Ni film on MgO(100) substrate". In: *Journal of Physics D: Applied Physics* 35 (2002), p. 1864 (cit. on p. 84).

[79] P. Bruno and J. P. Renard. "Magnetic surface anisotropy of transition metal ultrathin films". In: *Applied Physics A – Solids and Surfaces* 49 (1989), pp. 499–506 (cit. on p. 86).

[80] P. Poulopoulos, J. Lindner, M. Farle, and K. Baberschke. "Changes of magnetic anisotropy due to roughness: a quantitative scanning tunneling microscopy study on Ni/Cu(001)". In: *Surface Science* 437 (1999), pp. 277–284 (cit. on pp. 86, 99, 100).

[81] U. Bauer, M. Przybylski, J. Kirschner, and G. S. D. Beach. "Magnetoelectric Charge Trap Memory". In: *Nano Letters* 12 (2012), pp. 1437–1442 (cit. on p. 89).

[82] S. S. P. Parkin, C. Kaiser, A. Panchula, P. M. Rice, B. Hughes, M. Samant, and S.-H. Yang. "Giant tunnelling magnetoresistance at room temperature with MgO(100) tunnel barriers." In: *Nature Materials* 3 (2004), pp. 862–867 (cit. on p. 93).

[83] W. H. Butler, X.-G. Zhang, T. C. Schulthess, and J. M. MacLaren. "Spin-dependent tunneling conductance of Fe/MgO/Fe sandwiches". In: *Physical Review B* 63 (2001), p. 054416 (cit. on p. 93).

[84] J. Shen, J. Giergiel, and J. Kirschner. "Growth and morphology of Ni/Cu(100) ultrathin films: An in situ study using scanning tunneling microscopy". In: *Physical Review B* 52 (1995), pp. 8454–8460 (cit. on p. 99).

[85] R. Ramchal, A. K. Schmid, M. Farle, and H. Poppa. "Magnetic domains and domain-wall structure in Ni/Cu(001) films imaged by spin-polarized low-energy electron microscopy". In: *Physical Review B* 68 (2003), p. 054418 (cit. on p. 100).

[86] R. Smoluchowski. "Anisotropy of the Electronic Work Function of Metals". In: *Physical Review* 60 (1941), pp. 661–674 (cit. on p. 104).

[87] N. Lang. "Self-consistent properties of the electron distribution at a metal surface". In: *Solid State Communications* 7 (1969), pp. 1047–1050 (cit. on p. 104).

[88] M. W. Finnis and V. Heine. "Theory of lattice contraction at aluminium surfaces". In: *Journal of Physics F: Metal Physics* 4 (1974), p. L37 (cit. on p. 104).

[89] J.-M. Albina, C. Elsässer, J. Weissmüller, P. Gumbsch, and Y. Umeno. "*Ab initio* investigation of surface stress response to charging of transition and noble metals". In: *Physical Review B* 85 (Mar. 2012), p. 125118 (cit. on p. 104).

[90] M. Drechsler. In: *Zeitschrift für Elektrochemie* (1957), p. 48 (cit. on p. 105).

[91] G. Aers and J. Inglesfield. "Electric field and Ag(001) surface electronic structure". In: *Surface Science* 217 (1989), pp. 367–383 (cit. on pp. 105, 107).

[92] F. Weigend, F. Evers, and J. Weissmüller. "Structural relaxation in charged metal surfaces and cluster ions". In: *Small* 2 (2006), pp. 1497–1503 (cit. on p. 105).

[93] R. Nichols, T. Nouar, C. Lucas, W. Haiss, and W. Hofer. "Surface relaxation and surface stress of Au(111)". In: *Surface Science* 513 (2002), pp. 263–271 (cit. on pp. 105, 106).

[94] Y. Umeno, C. Elsässer, B. Meyer, P. Gumbsch, and Weissmüller. "Reversible relaxation at charged metal surfaces: An ab initio study". In: *EPL (Europhysics Letters)* 84 (2008), p. 13002 (cit. on p. 105).

[95] R. P. Feynman. "Forces in Molecules". In: *Physical Review* 56 (1939), pp. 340–343 (cit. on p. 106).

Acknowledgments

- First of all, I would like to mention Prof. Wulf Wulfhekel who gave me the possibility to carry out this thesis: thank you for the guidance throughout the last years, the freedom that you allow your students and your relaxed manner in all situtations. I really would like to thank Prof. Georg Weiß for willingly agreeing to be the second referee.

- I am really grateful for the theoretical support from the groups of Arthur Ernst, MPI für Mikrostrukturphysik in Halle, and Ingrid Mertig, Martin-Luther-Universität Halle.

- I would like to express my gratitude to Prof Y. Suzuki and his group from University of Osaka for welcoming me. I am much obliged to KHYS for the extensive financial support of my three-month internship in Japan.

- I really enjoyed my stays in Margrit Hanbücken's group from CNRS Marseille and especially, I would like to thank Eric for his kindness and his efforts.

- I wish to acknowledge the supervision from Toyo when we were working at the RT STM, writing papers and on conferences.

- I thank Tobias, with whom I shared my office during my Diploma and PhD, for the positive and relaxed atmosphere during all these years.

- I would like to thank Toshio for teaching me the Japanese style and I really appreciate that he always gave me his frank opinion.

- I would like to thank Stefan S. and Timofey for helping me with the experiment at the beginning of my work.

- I really appreciated to work together with Michael on the difficulties of depositing single molecules. I learned a lot when I was working together with Rien on MEC in Fe/Cu and I would like to say thank you. I would like to thank Moritz for his contribution to the work on Fe/Ni: together we survived numerous night shifts and the troubles of an always dirty surface. I also thank Max for the time that we spent together working on Fe/Ni.

- I am truly thankful to Fred and Shigeru for the intensive assistance during my stay in Osaka, not only on the scientific level.

- Finally, I would like to thank Katja, Moritz, Max and Toyo for proof-reading.

Experimental Condensed Matter Physics
(ISSN 2191-9925)

Herausgeber
Physikalisches Institut

Prof. Dr. Hilbert von Löhneysen
Prof. Dr. Alexey Ustinov
Prof. Dr. Georg Weiß
Prof. Dr. Wulf Wulfhekel

Die Bände sind unter www.ksp.kit.edu als PDF frei verfügbar
oder als Druckausgabe bestellbar.

Band 1   Alexey Feofanov
Experiments on flux qubits with pi-shifters. 2011
ISBN 978-3-86644-644-1

Band 2   Stefan Schmaus
Spintronics with individual metal-organic molecules. 2011
ISBN 978-3-86644-649-6

Band 3   Marc Müller
Elektrischer Leitwert von magnetostriktiven Dy-Nanokontakten. 2011
ISBN 978-3-86644-726-4

Band 4   Torben Peichl
**Einfluss mechanischer Deformation auf atomare Tunnelsysteme –
untersucht mit Josephson Phasen-Qubits.** 2012
ISBN 978-3-86644-837-7

Band 5   Dominik Stöffler
**Herstellung dünner metallischer Brücken durch Elektromigration und
Charakterisierung mit Rastersondentechniken.** 2012
ISBN 978-3-86644-843-8

Band 6   Tihomir Tomanic
**Untersuchung des elektronischen Oberflächenzustands von Ag-Inseln
auf supraleitendem Niob (110).** 2012
ISBN 978-3-86644-898-8

Band 7   Lukas Gerhard
Magnetoelectric coupling at metal surfaces. 2013
ISBN 978-3-7315-0063-6